Wings and Tracks

By
David Childs

The Caxton Printers, Ltd.
Caldwell, Idaho
2009

Cover photo:

Original caption, "Luke Field, Ariz.—The silver wings of a fighter pilot were pinned on Lt. A David Childs, of Arlington, Ore., by his wife when he graduated from this AAF Training Command flying school on Feb. 8, 1944. (AAF Training Command Photo)" *

*From photo sent to home address, courtesy of Command. February 1944

Copyright © 2009 by David Childs

ISBN   978-0-615-30497-7 $18.75

Jeanne Hillis lives in The Dalles, Oregon; has many interests – artist and member of the International Women's Pilot Organization, the Ninety-Nines. Jeanne stopped piloting and donated her four-place powerful bird to a world-wide charity foundation. Diminutive Jeanne reasoned when one's age and weight numbers near or meet, one might modify a little.

Acknowledgements

Miss Martha Hill was friendly and effective in teaching Arlington High School students home economics, bookkeeping, and typing. She also was advisor to the fourteen member sophomore class. Total AHS enrollment was sixty-six in 1938-39, my sophomore year.

School policy required all sophomores to take typing I. Miss Hill's skills allowed us some tolerable informality. This gave us another benefit, the typing room's magnificent view. We looked north from three stories above the schoolyard which in turn was two-hundred concrete steps above the street below. Our vantage viewing was high and our scope wide.

The typing room was long and narrow, formerly being the stage for both high school and grade school productions. Two rows of desks set in slight stagger allowed all intense learners to catch restorative moments by looking out. From the west, the distant distraction was "The-Good-No-Hills;" to the east, it was "The Horse-Heaven Hills." Both had Native American and settler American families sparsely using this scenic backdrop.

The village of Roosevelt, Washington (Teddy), population under one-hundred, was miles closer. It served its "Great Northern" railroad bordering the north side of the mighty Columbia River. Arlington, population six-hundred, on the Oregon shore had its "Union Pacific" railroad, a transcontinental highway, US 30, and was hub for two branch railroad lines serving cropping, grazing, and timber lands to the south.

In typing I class, not much thought went for transcontinental railroad or our international river connections; however the interstate ferry was visual entertainment. We caught glimpses of the man with his little mules crossing with his sailboat; or the ferry hauling depression day vehicles and Native American Indians with horse drawn buggies. One native told me, in pre-ferry days, they swam the works to an island, stayed overnight, then continued with rested teams to the other shore next day.

High water, low water, frozen river, ice jams, wind and wave hazarded ferry crossings, highway and train caused fires from tossed cigarette butts, and sparks and hot boxes; all were spectacular.

In retrospect, typing class I, its teacher, subject, and classmates, molded bits of me, but my view and point of view were influenced equally through the window. Thank you all.

Inadvertently, we got exposed to geology, climatology, economics, sociology, transportation, energy, human relations, and with an airfoil folded sheet of paper, an open window in May, gravity and aerodynamics.

I wrote of the library, but not the telephone exchange, always operating "24-7," as was most of Arlington's travel related businesses. Arlington ferry linked the grain growing communities with goods and services.

Union Pacific Railroad tracks and depot were close below just over the end of the add-on gymnasium. US Highway 30 turned right onto main street a block up from the depot and tracks and in four and a half blocks displayed eight gas stations, two hotels with dining rooms, three garages, two pastimes, eight restaurants, two grocery stores, two churches, two farm machinery dealers, two barber shops, one secondhand store, the laundry, and two grain buyers tied to warehouses. Apartments were above the closed bank. The hardware store, drugstore, Rio Theatre, and Red and White grocery were lined across from the bank. Banking was not visible; it must have been worked out, maybe in the Masonic hall over a closed furniture store.

Nine year old "Young Doc" practiced trumpet in their apartment above "Old Doc" Severinsen's dental office with its ancient chair and foot-powered drill. The harness and shoe cobbler's shop, men's clothing store and women's fashion store, and post office were down the street. Mail came by train four times a day. Greyhound depot was a covered part of the Oasis cafe. Milking, pumping gas, delivery, shoeshine, cleaning, waiting table — everyone worked or could get work.

Maude sold handmade, beaded, white or tan, buckskin moccasins and gauntlets from the steps of the post office. She tanned the hides at their camp way out beyond the two auto courts; past the football/baseball field, and one of three dairies. Her family's semi-permanent camp consisted of a tepee and a wooden building made from Columbia River flotsam. The camp was among large sagebrush used in smoking stretched deer hides. I jogged by their camp after school during track season.

Arlington's municipal airport was high on the east bluff.

Acknowledgement to both old and new Arlington are easy and important to me.

My Dad, L.W. Childs, and Mom, Cavy E. Weatherford Childs, high school graduation gifts were a desire to read and to learn plus a Royal portable typewriter. I was pointing for a Piper Cub for college graduation.

More thanks go to a lifetime of fine neighbors, teachers, flying and POW mates, some not mentioned, all are remembered.

Encouragement over these last few years has really boosted our morale. Writing critiques from an English teacher, an editor, several world travelers, farmers, politicians, family and community college teaching talent, twenty years ago and again recently, gave a real boost.

In December and January 2006 and 07, six weeks with the staff at American Lake Blind Rehab Center for Veterans picked me up as much or more than Miss Hill and our typing class sixty-eight years before. Royal portable? What is that? Regaining lost keyboard and improved PC skills was the gift enabling my finishing an effort started in 1984.

Peggy and daughter Kit with heroic effort have commuted by email with affect or maybe it was effect. Kit and Chris and spouses, Denise and Curt, have given yeoman support and tolerance for years. Here, I use an often heard L.W. quote, "Thank you until you are better paid."

Finally, I thank our nation for the way we pulled together for those war and post war years. They started with Liberty Bonds in WWI. Then we joined again in WWII with bond drives and buying saving stamps. A twenty-five dollar war bond cost $18.75. Trying to figure a cover price for the book seemed tough until the thought of a 1944 war bond seemed to fit. Thank you.

*Peggy and I dedicate this book
in memory of comrades lost,
and to those who went to war and to those at home;
and especially to our parents and their grand and great
grandchildren.*

Table of Contents

Chapter 1: Series of Events 1

Chapter 2: Fender 9

Chapter 3: Dad 'n' Me and the Model T 14

Chapter 4: Kaiserslautern 19

Chapter 5: Anne and Louise 24

Chapter 6: Gold Stock to Wedding Band 33

Chapter 7: Tracks for Sure 42

Chapter 8: Peggy and David 48

Chapter 9: Peggy 56

Chapter 10: Stalag IV-B 59

Chapter 11: The Ward 66

Chapter 12: Jene and I 75

Chapter 13: Had a Beer 85

Chapter 14: Steps and Sugar Cubes 91

Chapter 15: February 21 1945 96

Chapter 16: Nurnberg 103

Chapter 17: Water Bottle 111

Chapter 18: White Bread 117

Chapter 19: Barracks Bag 125

Chapter 20: Ray-Bans & Hot Dog 130

Chapter 21: DFC and Bath in Tub 135

Chapter 22: Home June 1945 141

Chapter 23: Getting Acquainted 1945-46 148

Chapter 24: Progress 159

Chapter 25: Looking Ahead / Living Now 168

Wings and Tracks Part 2 197
Chapter 26: Goose Pits and a Different Wheat Field Set-Up 198
Chapter 27: Flying Again 203
Chapter 28: Friends and Memories 209
Chapter 29: London Maidstone 215
Chapter 30: The Landing 84 219
Chapter 31: Combat 224
Chapter 32: Rennes to Paris 233
Chapter 33: Dayton Ohio 1960: Our Billy and Our General 245
Chapter 34: Living 250
Chapter 34: Somber 259
Chapter 36: It's Tough, Lighten Up 265
Chapter 37: Washed Vision 270
Chapter 38: Joy & Maudlin 276
Chapter 39: Quest 282
Chapter 40: Vinegar and Wine 92 288
Chapter 41: English Channel D Day + 50 297
Chapter 42: Liesel 303
Chapter 43: Next Day 314
Chapter 44: Rhineland to Normandy 319
Chapter 45: Beatrice's Normandy 329
Epilogue: Connecting Then to Now 341

Chapter 1:
Series of Events

Nowadays, on November 17th, I wear a necktie with diagonal rows of miniature American flags on a dark blue background. The tie was given to me in 1976 to honor our nation's two hundredth birthday.

During wheat harvest in the summer of 1930, Dad hired me to ride my horse to keep cattle away from sacks of wheat until it was hauled. I was seven and my horse was twenty-five. The job was six days; fifty cents a day. Dad took me with him to town. I bought three small left-over Fourth of July American flags from Mr. B, city recorder, for a dollar and fifty cents.

My wife Peggy's Testament, a childhood gift from her grandmother, was in my shirt pocket, as it had been on all my missions, this day November 17, 1944.

We had been grounded by rain and fog. Morning hit us with cold and an inch of snow. I added long johns and a GI pullover sweater before heading to the squadron ready room. Clearing and sun breaks were key words in the weather forecast.

Just maybe we can get a mission off and catch Jerry moving before he gets stuff out of sight. I change my goggles from clear to an anti-glare lens for better vision when looking down into snow and shadows. Armed reconnaissance, three flights of four, plus a two ship element--fourteen ships: It looked like the mission might go. Weather moves across France from us to the enemy. Our meteorologists calculate weather for ground targets as it comes in by phone, radio, teletype, and mission debriefings.

This sweep route covers roads, railroads, and road targets and anti aircraft guns. We have flown several missions just a few miles north. These were focused on tunnels and bridges, incidental duels with anti-aircraft batteries, races with locomotives for tunnels. Today, fresh snow and sunshine heightened anticipation. "We will find them."

Ground crews are ready with their P-47 Thunderbolt fighters serviced, armed, and tuned. Pilot briefing is short. Most surviving original squadron pilots have finished their tours, about one hundred missions, and rotated home. Some have come back as squadron leaders or moved up to Group.

"Weather" estimates an hour before takeoff; the target area is still socked in. We know through experience "weather's" predictions are good. We count on their accuracy. For example, one mission we took off wrestling tree tops and scraping cloud bottoms at the same time during join up. We broke out at ten thousand feet, continued to a pin point target, dive bombed it successfully and got back to our strip, low on fuel, yet in good enough weather for landing just as they had forecast.

Fourteen Pratt and Whitney R-2800 engines are fully checked and warm; bombs loaded. Fourteen restless pilots exchange their billfolds for escape kits from numbered small blue bags held open by the intelligence sergeant. He tugs the draw string and hangs each on its nail. Large white numbers identify each bag. Mine is number 46.

Escape kits are about six inches square and a little more than an inch thick. They fit in a patch pocket on the right leg of our flying suits. They are fully packed with survival and escape gear; cloth map, compass, chocolate power bar, water purifier tablets and more.

Word comes down "start engine" time is probably forty minutes away; considering the current low ceiling, maybe longer. Time enough to feed his pilots, our squadron CO gives us the choice, early lunch or chow when we return. Someone rumors, "Steak." Three jeep loads head for the château dining hall.

Real surprise, beef steak is speared from a stack on back of the stove, already cooked done and brown, no juice. I move up in line holding my tin plate at arm's length. I lean ready to lunge because the flamboyant server is arching steaks at eye level across the serving table between me and the stove. My steak dangles by one fork tine and I am not going to miss. I bag it just when the message comes, "Start engines in ten minutes."

The weather window is struggling to open; we help "weather" similar to the way a high jump fan flexes a leg watching jumps. I slap my steak between two slices of GI bread and hop, chewing, into a jeep heading for the flight line. Chute on, still chewing, I get a boost into the cockpit. I swallow the last dry bite of bread and beef as I buckle into shoulder harness and seat belt.

We take off by twos. I am red three, my wing man is red four. We are the second pair off; we are red flight's element.

Armed reconnaissance missions are flown between one and two thousand feet above ground. This is a good height for both sighting and starting attacks. The Lieutenant leading the squadron is one of the first replacement pilots, joining two months before us. I have flown several times with him including a recent two ship weather mission, to check conditions near enemy targets and radio back information for the group. I like him and feel comfortable.

My wingman is a first Lieutenant who transferred from the Army to the Aviation Cadet Program in his military grade. He and five others came to us and P-47s when their P-51 fighter bomber group became a photo recon outfit after their group was so severely clobbered on a mission that higher-ups disbanded the surviving remnants and placed them with ours and other clobbering Thunderbolt squadrons. I have been leading elements for six weeks and flights in a couple situations. I have had silver bars a week.

Flight leaders carry target maps; rest of us fly both offense and defense, turn into enemy fighter attackers although our specialty is ground targets and in combat protect the unit so leaders may concentrate on radio, navigation, and attack. Missions flying from our airstrip, A-79 east of Reims, cross miles of desolate WWI battle fields. We pass near Verdun. In two days, our unit is moving to Etain east of Verdun. We stay south of Metz, a hot spot with Patton's Third Army battling for a breakthrough. We have struck along here and to the east several times.

Today, we spread into flights of four ships abreast plus one two-ship element in open battle formation, sweeping over an area two to four miles wide. We are about at the turn in our sweep. My wingman and I are on the right of the spread of ships, in a wide turn to the left. The outside flight is doing a crossover behind us when I spot two locomotives, caught like kids smoking behind the barn, only there is no smoke; just side by side under a bridge on a branch off the mainline.

I call Red Leader, getting his OK to attack. He says, "Go ahead, catch us on the swing around."

I call my wingman, Red Four, saying, "I'll get the one on the left. You get the one on the right."

I get good concentration; hits through the cab into the firebox and boiler from the rear and pull up sharply in a climbing turn to the left. I look over the targets. My locomotive is blowing steam. My wingman is coming across a little below me, appearing to be out of position to have made the strafing attack. I am back to initial attack position. I size up the situation, two locomotives side by side on double track under a low bridge. The roadway over the bridge comes along the hillside paralleling the RR tracks, crosses the narrow valley straight into a small village. An approach other than the previous will have steep hillside or village difficulties. Our mission is to disrupt supply to their troops fighting Third Army. The squadron plan is to hit hard, keep moving and join up in group position. We need to keep the fighter sweep-reconnaissance moving. A mainline marshalling yard is down track a mile. Our leader passed it up to keep the sweep going. We catch targets that should not roll again.

One is damaged and spouting steam. Both locomotives are on the rails in condition to tow or be towed. I call Red Four saying, "Cover me. I'm going in to get the other one."

I am a little abrupt for losing time, but more so for his failing to get the second locomotive. It appears to still be fat and happy. I saw no ground fire the first pass, however, on this second go, I up the speed trying to tip the odds in my favor. I arm the two five hundred pound bombs. I aim to get the works, lay them in against a backstop target of side abutments, short overhead bridge, both locomotives, and rails, all in one exploding tangle.

It is "down the wire," straight in, with four to five second delay fuses...I release the bombs and shove full throttle, push the nose down as the lightened Jug clears the target, charges full bore, like giving reins to my saddle pony jumping a wash in sage brush heading after a critter, skims low, on the deck, going upstream alongside the creek. The 362nd calls this "cabbage cutting."

I feel energized speeding up the little valley and bank hard at first side canyon. I look quick. Glimpse a pair of 20mm guns a lariat's throw below my vertical banked Thunderbolt's cockpit. Two helmeted gunners stare, mouths open, half faces; they look rigid as a scene in Madame Tussaud's London wax museum, their pivoting necks not fast enough to catch up at six hundred fifty feet each second.

Two hits in my plane, "thump thump." "Stay low. Strikes must be from another anti-aircraft battery."

The hits feel quick like two cadence beats on the bass drum starting the music in a marching band or maybe crossing ruts in the snow while speeding down schoolhouse hill on my Montgomery Ward Flyer.

Four or five seconds since I released the bombs; we are somewhat even. Instruments check OK. Engine? Sounds perfect. My right hand switches on full oxygen; thin smoke seeps around my right arm. Bright sunshine, fresh snow outside. Skimming across slightly rising rolling ridges, trees in gullies; too fast to belly in, too low to jump.

Black smoke floods cockpit. Shove goggles up. The sixth second, right thumb hits canopy switch a flick to vent. In a milli-second, ignition. Right hand thrusts straight into a blow-torch flame, yanks emergency jettison-canopy release.

I had practiced quick exits when I cut the ignition switch after missions. It took a quarterback's long count to unhook harness, seat belt, oxygen and radio and to scramble out; this time only, "Hike." Practice wasn't in plus-four-hundred mile per hour inferno.

I'm a spectator. Each hand is individually automatic. Left jerks the stick back. Right releases the shoulder harness and seat belt latch. Feet thrust down; legs launch body up. Slipstream grabs me by head, yanks me from the Thunderbolt like a drogue chute pulls out a packed parachute. Body snaps like a bull whip, head to toes. I spin in her wing-tip vortices, hands at knees; did the departing Jug pull me up in her dying arch? "Bless her."

My brain cuts in. "D Ring!"

My extended arm pulls against centrifugal force. Hand grabs and jerks--eleventh, twelfth, or thirteenth; it was the right second.

The rip cord activates the sequence in a blink. Nylon snaps and jams jaw and spine; violently jerks and stretches knees, ankles, toes, and shoe laces. The Jug blows up in a red-gold fireball dead ahead. Tree tops change my focus to down and close. Elation applauds my short straight descent. My feet jam through three inches of fresh snow. I hit squarely in one track of a two-track forest road on a slight slope. My rump slams past my knees and thuds down on the low rise between the snow covered ruts. I think, "Thank you God," then of my condition and of evading.

I am down in forest hills south of the Moselle, west of the Rhine and east of Saarbrucken. It takes effort to balance my body over my knees. They act as if on stilts with ball joints below the step. I delve for comprehension. My left eye is dim; my right is seeing but swelling fast. I attempt to scoop snow on my face. I can't. My leather gauntlets are burned, encasing each hand tightly in fire shrunken leather.

My chute harness has a pocket with a large easy to move zipper. I keep a marine K-bar knife inside, just in case. Using my arms and hands like tongs, I slide the zipper open. Fumbling with the knife, I slit the back of the left gauntlet; and slit each finger stall. The light knit wool gloves inside the gauntlets show torn finger tips but protect my hand and speeds the tedious skinning out of my encased fingers.

I am wearing T-shirt, shorts, long-john drawers, "greens" wool shirt and trousers, knit long-sleeved sweater, the leather gauntlets with wool liners, and twill flight suit. My trousers are tucked into buckled and laced combat boots, paratrooper style.

Will I get better, quickly shake it off? Will my legs function? Can I get away from the site and somehow lose myself and tracks?

White chute over snow, tree top height, same time the aircraft explodes into hillside; my jump probably is not seen by either friendly or enemy eyes. "How bad are my eyes?"

I glimpse my hands; both thumb nails are black as though smashed with a hammer. I raise a hand up to my forehead. It feels like leather. I think, "How can my leather flying helmet stay on?"

I reach with both hands for chin strap and ear flaps. There are none. There is no helmet. I feel hide. I do a hasty inventory. My limbs seem to be working but sockets are loose. I check for my escape kit. Both it and the leg pocket holding it are ripped out and gone. My feet slip loose in my boots and the laces are slack. My nose and mouth seem OK. I left ignoring the oxygen mask and radio plugs.

I slip the knit wool glove liners back on my hands. Some of the glove's fingers are holed at the tip and burned. The chute is lying over wild rose bushes and shows rows of saucer-like blackened holes, like

Series of Events

unfolded cutouts. Those extra vents may have eased the impact of the chute opening or swinging me with a vicious body slam into the ground.

The snow is unbroken; not a track anywhere. No use to bury the chute. Everything I did would be written in the snow. My dog tags are missing from my neck. A string holding the jewelry chain ends must have burned through. My forehead and eyelids are burned severely.

Need get away from this forest track. I walk uphill, one step at a time through park-like groomed forest--slow going--balance body above loose knee sockets. Eyes closing. Press snow on right eye.

I hear young voices moving ahead--Glimpse hurrying figures through an opening--Feel like an elk partially concealed in timber, watching a hunter go by.

I slip into the end of a brush pile, hunkering down. They are walking the ridge in an opening. Their voices fade to my left. I need to get moving. Which way? I try to think clearly. How bad am I hurt? Can I make it on my own? They are probably civilians. I remember vaguely, "Avoid citizens. In Germany, your chances are better with the military."

Decision time ends quickly with the sound of a motor vehicle coming up the forest track. Engine means military. It stops down near my parachute.

Time to move. I try to get up. I am stiff. My knees are tightening; my tendons don't work. My jaw will hardly move and my sight is dim. The options are all centered right on this spot.

My ears and senses tell me I will be in plain view soon after they start on my track and look up. I conclude I have only one line of defense. That is hold my captors' respect. I can see maybe thirty yards. By the time I can stand, I walk measured and very clumsily.

Their low voices seem cautious. They move uphill. I am spotted. I don't want to make any excuse for them to fire. An accented command comes up the hill, "Hands up."

Dim walking figures show in my curtained vision. I raise both arms, hands pointing toward my eyes; a gesture about halfway between a referee's signal for a touchdown and a country preacher's blessing of the Sunday offering.

Snow muffles their steps--they come on--spread three abreast. The leader stops close in front of me. He says, "For you the war is over. Will you have a cigarette?"

I expected his statement's beginning words but totally miss the question. I reach for the upper left pocket of my flight suit where I carry an unopened pack. It, too, has ripped away. I say "I'm sorry but my cigarettes are gone."

He replies, "No, would you like a cigarette?"

I respond, "No, thank you."

He asks. "Are you armed?"

I point to inside my pullover. He removes the large marine knife in its scabbard, saying, "Souvenir."

He reaches into my chest pencil pocket, retrieving a small pen knife. He hands it to another, saying, "Ein more souvenir."

I didn't ask him to look for my favored ranch stock knife which I discovered had torn out through the bottom of my trouser pocket back yonder.

I give him my name, rank, and serial number. I tell him my dog tags are gone. He asks how many were in my airplane. I tell him according to the Geneva Convention, I am only permitted to give my name, rank, and serial number.

They steady me, one at each elbow. We slowly start down the hill. The car is a small sedan. At this time, we are in bright sunshine. The Sergeant places me with my rump over the wheel, me leaning against the left front fender and says, "You will stay while we search for the others."

I am left with one guard. I am anxious to get away from the vehicle. It is a likely target. I feel bravado in standing my ground. It is a small accomplishment, but I am satisfied that it is right to not tell them anything, although I want to say, "It's a single seat plane. I was alone. I need medical aid."

I had seen rabbits burned badly in grass fires. I knew it was a fine line between trying to save them, or putting the poor devil out of his misery. I feel the need to be strong. I know I am very fortunate to be alive, but I am seeing less and less. I wonder if I will be able to stand up or walk for much longer, even with aid.

Chapter 2:
Fender

November 17, 1944
A Glen in an Alsace Lorraine National Forest

I feel apprehensive. Waiting is not my strength. I change foot position. Not a sound. No birds. No aircraft engines, no voices. Sun must be obscured, isn't shining on my bare head. My squadron should be nearing base. Leaning propped up against a small German vehicle's fender is not my notion of progress. I reason, the Sergeant is doing his job, determining the situation according to his duties. I am doing mine according to my duty also. I think, "It may be a while before we are both concerned about my wellbeing."

The vehicle and I are arranged in this snow covered scene like a posed pulp magazine cover.

The German Sergeant had carefully asked, "How many others?"

I had already told him, "I am only permitted to give my name, rank, and serial number."

He instructs me, "You will stay. We will look for the others."

He leaves me alone with the third soldier, probably the driver. I can make out the rifle in his hand with the butt grounded. Neither I nor my guard speak. I feel less tense as the minutes go by because of the Feldwebel's civility. I guess capture was about noon. I relax a little with each waning minute, especially when the cloud cover begins to fill overhead. I don't turn or attempt to move. My systems, both thought and physical, are nearly in vertical repose like a ground tied horse. My right eye can see down to foot tracks and my boots.

The two searchers return. They have been gone less than an hour. My crashed plane must be fairly close. They say nothing to me, but talk briefly. We load into the car, the Sergeant beside the driver. I sit behind the driver. The vehicle is smaller than our Baton Rouge Ford 60.

My knees and jaw are stiffening, have pain in my left little finger. I worry about my swollen forehead and closing left eye. I can still see the men's figures; hear their voices in discussion, but awareness of outside is limited to light and shadow and the engine and road noises. We turn, crossing what might be a railroad track. I try to see if there are tangled locomotives but fail. I speak to the Sergeant, questioning, "Herr Doktor und staff?"

I think I am not really bad but in much worse condition than I first reckoned. The Sergeant indicates he understands but gives no encouragement.

We turn onto a smoother road and drive about two miles into a village. The car is parked. I am walked over cobblestones, slightly downhill, held at each arm. The Sergeant protects me as we move along. I feel people closing in and "jawing" me. The Sergeant speaks sharply. Those crowding back off. I am thankful. His tone is firm, means business and has neither swagger nor plea. It is just plain reassuring. If I interpret his meaning correctly, he says, "This man is my prisoner, and I intend to keep him from harm. It is best you back off."

I am turned right into a building then up a long flight of stairs, not as steep as in our 1883 ranch house, but a few more steps. They give me time but no help. It is steady, one step, next step, next step. Both legs lift, working albeit slowly. I think, "Do it! Hold your head up."

I am seeing better than I let on, but my view is curtained, nothing clear. My left eye is swollen shut. We move from the top landing into a quieted room about the size of the Arlington city hall's meeting room. Low talking comes from dim figures. People are seated along the walls. They take me to the center of the room. I am glad to accept a chair.

I feel debilitated, maybe even worse than when I ran one ski down a badger's hole; self-made skis, broken rib--Doctor Wilhelm made two twenty-five mile house calls, used yards of tape; days in bed; Seventh grade, Olex country school.

The chair is wooden. I sit a few moments before another chair is placed opposite me. It is like chairs I remember.

I am cloudy about the details because a person speaking conversational English sits down, moving his chair slightly. His action perks me up. I darn near grin. His chair creaks familiarly.

These chairs are the same as in our city hall at home. Their seats and backs are made with little slats held in place by grooved wooden frames. I strengthen.

I and my interrogator are isolated in the room's center. He asks a few questions. I stay with name, rank, and serial number.

The chair is a powerful benefit to me. I sat in a like chair in Arlington City Hall for high school band practice. Our city library rimmed the walls of the room. The librarian, Robbie, was one of the most kind, practical and unflappable people I ever knew. However, the books that tall Tennessean encouraged me to read did not teach me all of life's realities. Robbie introduced me to Richard Halliburton and his daring travels and adventures in his open cockpit bi-winged, "Flying Carpet." Through the magic of books, I made great discoveries. With daring, fortitude, and planning, anything was attainable in my fledgling years.

Our band leader conductor, Mr. Simpson, routinely put me on the spot. He, even at times, was genuinely incited to being excited by my bass drum bungles. I was the Herkimer of bass drummers that freshman year. Doc Severinsen was our band's ten year old first trumpet.

The interrogation chair is a bonus. The home town feel in this traumatic circumstance is a gift.

Few questions are asked. It is more like me being on display, and they waiting for further direction. I think it also is to ensure my personal protection until instructions arrive. For that I am grateful.

Soon I am led to the top of the stairs. I am unable to take the first step down until a person helps. The descent is faltering, one step at a time.

Outside across the street, we walk downhill. I am guided into a small building. I dimly see a doorway framed under a one story gable roof. The room is rectangular with a single width bed just inside on the left. They indicate for me to lie down.

We had taken off about ten hundred hours. I was downed about eleven hundred hours and tracked down close to noon. Now it is a couple hours later. I am losing the adrenalin. My system is backing off. I lay down on the straw filled mattress. I surmise, "So far, so good; Third Army is about to take Metz; tip of my left little finger hurts like blazes."

I slip into a finger throbbing, hazy catnap. My guard says nothing. I am soon awakened by a new agitation. It has been at least three hours since I climbed into the P-47, eating a dry steak sandwich. It interrupted my preflight proceedings. I might have missed my take-off prayer. Now I know I skipped the "scare pee" usually done on the walk-around before climbing into the cockpit.

These omissions disturb me. How could I have been so hurried? I think back: Barker was leading the squadron. I led his element. I spotted the two locomotives side by side under the overpass on the west side of the village. My thoughts of the mission, 'good strafing with a tall spouting steamer on the first pass and good bomb release in the groove with second pass. Lucky 'aak aak' gunner; smoke and flame. I drift off. Soon feel a containing bladder along with the paining little finger. Without seeing the guard, I ask, "Urinate?"

He replies, "Was iz?"

I point at my groin, sort of hissing, "Sss sss," like I think a relaxing stream might sound relieving a twenty-one year-old's bladder.

The guard responds nonchalantly, "Ja, ja," and something I hope means SOON!

In a few minutes, the door opens. The two guards talk a couple moments; one makes it clear I stand. I swing my feet to the floor. My knees are a little higher than my hips and nothing seems able to lift. I try and sink back down after rising as far as my arms and knuckles will push. I am sure I can do it but before I can try, the guards grab wrists with hands in what I call the "firemen's carry." City cousin Ed taught it to me when we were ten or eleven.

I understand and slide onto the clasped wrists. They sidle out the door. It is still light. People gather in, some quite close. The crowd sounds more neutral than earlier.

They carry me a few yards and stand me on a wooden walkway, then walk me a few steps to a railed footbridge. I hear running water below. People close in, some on the bridge. I am reassured with being able to stand up but must look as puzzled as I feel. A familiar voice comes forth with, "Ssss, sss."

The water trickling under the bridge, the appreciative crowd murmurs, and a successful communication combine to relax the constricting mechanism of my plumbing. What a relief! Cousin Ed would have been proud of me as I live one of his several joke titles. This one is so appropriate. "The Golden Stream" by I.P. Freely.

Ed also taught me a few words of German. In Portland, they taught foreign languages. In Olex seventh grade, we learned a little Chinook Jargon. Chinook would have been something like "HyU clatawa" roughly meaning, "Big Go."

The bridge scene, including standing, unzipping and zipping, guarantees parts of me work. Seeing boards tells me I have sight. My mouth grins, so it is OK, but my sight is fading.

That first impression of my forehead being my burned leather helmet concerns me a lot. They help me walk to the little car. As soon as we drive off, I again ask, "Herr Doktor und staff." This time they reassure me.

We drive uphill for several miles and stop in deeper snow. I see enough to discern a parking lot in front of a low military looking building. We walk along the outside to a side entrance. A huge man in white stands outside the door. I think, "My gosh, it's Herman Goering himself." I follow him in. "Goering" turns out to be a medical person. I sit down. The medic wraps my hands and wrists and bandages my head, leaving only a slit for my mouth.

It seems dark before we leave the aid station. I am carried out on a stretcher and thrust into the back of a military ambulance, my head to the rear. We drive up and down and up again, arriving, maybe thirty minutes later, to where the vehicle drives around corners like city streets, climbs uphill with smooth meshing gear shifting, and finally, backs into a parking place.

I am carried up stairs and around corners and placed onto what seems to be an operating table. My bandages are cut off. I feel my pants legs cut away. I am rolled onto my back. I look up into a bright light in what seems a totally white room. I see nothing I can focus on.

Chapter 3:
Dad 'n' Me and the Model T

I was ready and watching for Dad, listening, too, for the sound of the engine before he came around the street corner. I loved riding with him in the Model T car. It came quiet, top up, fine and dandy. It was Dad, for sure. He had hardly stopped before his black shoe came over the driver side door. The running board went down then up when he stepped to the driveway. I headed across the grass running, knowing we would give our best bear hugs. His wool overcoat and strong arms felt so good. I don't think I looked back.

My things were all in a box we put in the back seat. Dad was solid beside me.

Dad would pick me up Sundays to go visit Mom in Salem. I had to say goodbye twice on those days, once to Mom at the State Hospital and again to Dad at the home of the lady caring for me.

This was different. We were driving along 82nd but instead of going on toward Salem, we turned on Stark, toward Eastern Oregon.

When we passed the Twelve Mile House, Dad pointed it out, more as a reference. It was set back from the highway. He had a certain way of saying things, each with a connotation. In 1926, I don't think water was all they served twelve miles out of town.

Dad's family of eight brothers and four sisters, making fourteen with Grandmother and Grandfather, lived in North Carolina in the Blue Ridge Mountains. Many times in our ranch years I heard him say to others, "I came west at twenty-one in the spring of 1903, weighed 121 pounds, bought a ticket to Arlington, Oregon, got off the train, carried my suitcase and bedroll seventeen miles to Rock Creek where I had a friend working and a haying job waiting for me with Wilbur France."

I was three when Mom entered the State Hospital the second time. She would not return to us. Dad would raise their boy as they had planned, with love, encouragement, and opportunity. Several times I heard Dad tell friends, "Cavy went into the State Hospital first when David was eighteen months old. She came out after ten months. We bought our home on 84th Street, lived there only a year before she had to go back."

I'd gotten to ride with him in his other Model T, his delivery business truck with hard rubber tires. We'd say the names of Portland streets together: Stark, Ash, Burnside; and the schools, Buchman, Washington, Lincoln, Grant; and bridges, Morrison, Steel, Broadway.

Mom took me to the parade and opening ceremony of the Burnside Bridge. Soldiers carried flags, then men came by carrying tools like shovels and picks. Some pushed wheelbarrows and handcarts. We sat on the new cement curb on the downstream side on the approach span. My feet barely touched the roadway. That happy year abruptly ended with mother's return to the State Hospital.

I lived with Aunt Olive and Uncle Manly, Dad's brother, during Mom's first stay. Today, Dad was taking me to Heppner to live with his sister, Aunt Marion and Uncle Henry and their family. We didn't talk about being farther apart or sad things.

We stopped at the Vista House. The bathroom walls were smooth marble. Outside we looked over the bluff and up the Columbia from the walkway. I climbed up into the jitney by myself when we loaded up. It squealed the Model T backing-up sound when Dad backed out, then it did its smooth chuckle-along "galley galley" going around tree-shaded curves heading up the river.

We slowed by Multnomah Falls. Horse Tail Falls trailed in the pool close by the road. Dad named them all. Dad (Mom called him LW) talked and taught. They married in 1918. I was born in 1923. This September day, in 1926, was shortly after Dad's forty-fifth birthday.

We stopped at Bonneville fish hatchery. It would be a long day, but it wouldn't hurt to use it all. I stretched to see the big trout; Dad's tone was pleased when he pointed to them.

Back to the jitney and up the road, throttle pulled down. At Cascade Locks, Dad told about his ride on a sternwheeler through the canal and locks into the swift water below. He and his sidekick were going to the 1905 Lewis and Clark World's Fair in Portland. Later, he took me to see the huge Forestry Building left from the 1905 fair. The log and timber building was magnificent.

Dad eased "ole Lizzie" down the street through Hood River then on toward Mosier. It was preceded by some pretty crooked road. We stopped at the bottom of the hill for a visit to the cider factory. Golden cider-filled gallon jugs were beautifully lined up in an open trough of cold flowing spring water. It would be a couple of years before we visited Dad's old home in North Carolina on Three Mile Creek, between Buck and Doe hills, but he told me about grandpa's springhouse there. He said we'd visit there and see it.

We each had a glass of cider and bought a jug for Aunt Marion's family. Dad showed me some very fine trout when we walked back across the bridge to the flivver.

The restroom at the cider factory was my first experience with an outhouse. It was more interesting than the Vista House. We had left the ranch when I was nine months old, chamber-pot trained only.

We climbed East Mosier Hill to the overlook above Rowena. We looped around the flag pole. The Model T purred down the loops, and soon we were crossing Mill Creek into The Dalles.

The next summer, I stayed overnight at Mill Creek auto camp with my aunt and uncle and family. Six of us were in Uncle Henry's new Chrysler on our way from Heppner to Seaside. Dad met us at the beach.

Cider and curved steep roads encourage a rest stop. The Horn Pastime was a superior experience. Boy, what a sight and scent! We walked straight back me holding Dad's hand. I looked up at the horns and the stuffed animals. My nose was tilted only inches above the tall brass spittoons lined along the bar on my left. On the right, I couldn't quite see the cards on the tables, but I could see the players and cards held up to their faces. I got mixed signals again from Dad's tone of voice. I saw colored balls in leather net pockets. We never stopped until we got way back to the restroom. Then, back through the pool tables and the card games and along the brass rail with me looking up at freaks and trophies.

Dad took his time and carefully directed my eyes to the bald eagle, cougar, coyote, and deer, and the buffalo horns. There was a lot more. I sneezed. Bull Durham, cigars, chewing tobacco, and boot grease are penetrating.

On out of town, we stopped at Seufert's Cannery. Dad said, "Aunt Marion will enjoy a case of canned salmon."

We passed locks as we wheeled several barren miles on east to the village and Celilo Falls. We didn't stop long, but we looked across the canal to the platforms and dip nets. Dad said, "I sure would like to be an Indian just long enough to dip a couple of those salmon." His tone was only slightly envious. He had hunted for food in the Smokies with grandfather's muzzle loader, Dad's first gun.

I was as happy as I'd been for a while.

Soon Dad pointed across the river to Maryhill Castle. He said he'd seen Queen Marie and her tin soldiers at the Pacific International Livestock Show. A few years later, I heard him tell he paid fifty cents for a ticket, and the man in line in front of him paid two dollars. Dad enjoyed saying, "My seat was only two rows back of Queen Marie's special section. The fellow with the two-dollar seat was just one row in front of me!"

I barely remember crossing the John Day River. We drove through Arlington and out to my grandparent's home on Rock Creek. Dad's tone told me that Grandmother was a special kind of woman, tough and tender and to be loved and Grandfather was firm and to be respected.

Grandfather held me, "Cavy's boy," on his lap and showed me his agate watch fob and gold chain. I played with two footstools, carpeted wooden shotgun shell cases. I put one on top of the other so I could ride a pretend horse. I overheard Dad telling Aunt Marion, "W.W. [Grandfather] said all the grandchildren rode the footstools, but only David piled one on top of the other."

Coal oil lamps were new to me. It was daylight when I woke up. Dad helped me to dress quickly. The bathroom had only a tub. The kitchen sink had faucets. I already knew about outhouses from the day before. When I came back down the path, Dad was standing at the back door by the pitcher pump holding a teakettle. A wash pan was on the wash bench. LW's boy sure wasn't going to have dirty ears. When the scrubbing was done, I stood on a step while Dad showed me how to hold my hand over the pitcher part. He worked the handle while I sipped that very cold water. It was something never forgotten.

We ate homemade bread toast with churned butter and thick cream on oatmeal. We drank skim milk. Dad took my hand as we walked through the corral to the barn. We saw Mom's saddle mare, Goldie Gold Dust. I will always remember the pleasant aroma of alfalfa hay when Dad opened the barn door. The Horn Saloon experience had really made me aware of smells.

We loaded up quickly. It was to be my last hug from Grandfather. We hurried up the grade, Weatherford Road, going north from the creek ranch on the way to the hill ranch. Dad pulled the throttle down, adjusted the spark and said, "Ole Lizzie is sure going good."

George and Maude, two special big work mules, stood on the bunch grass hillside and followed us with eyes and ears as we drove by. I was holding their reins the next year. They needed little driving.

We would make it just like "Old Lizzie," doing our best. We doggone near reached the top before she needed to have the low gear pedal pushed in. Then she really pulled uphill and through the deep after-harvest dust.

We went to Arlington and up the river to Willow Creek and by Cecil. Then we passed Grandfather's 1870 ranch where they lived before he brought plows and "bob" wire to Shutler Flat; two poplar trees were at the site. We went by Morgan and around Horseshoe Bend. Dad was talking and teaching. He mentioned the Grady place and the horse and wagon drive in deep snow and Mom's train ride from Ione to Heppner and Doc Chick and how I was born in Heppner. He didn't talk about the move to Portland.

Now I am back at my birth town three months before my fourth birthday. I don't remember the goodbye but I sure do remember the next hello three months later when Dad came. I was sitting at the supper table in my highchair. Boy, he looked good! Dad had come to take me to Grandfather Weatherford's funeral in Arlington. Afterwards, I stayed on with Aunt Marion and her family until September 1927.

We were together from then on, me and Dad, with thoughts and prayers for mother. Getting an education and keeping clean ears were musts in my growing up with LW. He remembered his and Mom's plans for their son.

Chapter 4:
Kaiserslautern

I awake with a nurse attending me. I believe it is next morning. A urinal is placed in my hands. Although my hands are bandaged heavily, I manage with recall from cadet training. (Preflight situation, in Santa Anna, a bout with a virus; I was confined to bed, guzzled six measured quarts of water daily, measured by the numbers, in and out.)

Here, nurses address each other, "*Schwester Louise,*" sounded softly formal, "Swess'ta Louisa." The nurses, Anna and Louise, feed me one spoon bite at a time from a bowl. I cannot speak their language, cannot walk, and cannot see. The nurses attending me seem to be young, but perhaps a couple years older than I.

I try to keep track of days. I figure three meals a day. The days go by fast until I learn from an American Infantryman POW, they give snacks in the afternoon and evening. So, instead of five days, fifteen meals, I have been here three days, with five feedings a day. I am living "high on the hog" POW-wise. I feel stable, tired and recovering with limited appetite for these first days.

Somehow, I know I am in Kaiserslautern. The railroads near here, along with their tunnels and bridges, were often targets, although our mission on the 17th was an armed reconnaissance a few miles south. I am pretty well isolated by language and wrappings. I dream and ponder.

I recount to myself the events before and after as I know them. Analyzing is important to me, keeps my mind going. This mission, my forty-eighth, skirted south of Metz and south of Saarbrucken, a different route in hopes we might find fat targets. I led the element in Red Flight. Many of our senior flight and squadron leaders have completed their combat tours and are being sent home for 60 day leaves, or sent stateside into training command assignments. Replacements are given experience learning flight and squadron mission-leader rolls. Flying Red Three gives a close up-front view in handling a squadron formation.

We have one of the best leaders on this mission. My combat experience has taught me accurate delivery on the ground and in the air with eight 50 caliber machine guns; two 500 or 1000 lb. bombs; and on the job learning with four 5-inch rockets. I am gaining usable experience and could understand tactical ideas enhanced by my apprenticeship with my father in driving crawler tractors, pulling squadron-like multiple hitches of farm equipment in large open fields and through squeezes in close quarters.

I think about being alive and hope Peggy will know. I think of Dad and all his sacrifices and encouragement for me. They are strong people. Both have family and community support.

Our home communities buy war bonds, gather scrap metal, patch tires, ration gas, butter, shoes, eggs, sugar, meat, nylon stockings, and farm machinery.

Day one, I am in a hospital bed solid against a wall. My feet are covered, pillows are under my shoulders. Apparently, my bed is in a hallway between two ward doors. Immediately at my head is another bed occupied by a German soldier.

My head and neck are bandaged with only a small opening for my mouth and nose. The oxygen mask had protected my nose and mouth down to just under the chin. My hands are bandaged like a boxer's before putting on the gloves. My lower legs and left thigh are wrapped lighter than my right thigh which is heavily bandaged enough so the pajama leg is pretty well filled.

The nurses are OK. They hold the glass for me to drink. My care is firm and brisk. They sound pleasant enough as they talk to others.

The second day, I try to orient myself, wondering if I will get a full interrogation. I know very little. All of us received some training for interrogation. My first encounter starts after lunch. A raspy voiced interrogator begins with asking my name, rank, serial number, then continues in a sharper tone, asking about my unit and kind of airplane. I reply the instructed answer, "According to the Geneva Convention, I am permitted to give only my name, A David Childs, my rank, First Lieutenant, serial number 0767034."

The interrogator patiently listens while I answer his questions, repeating, "I am only permitted to say my name, my rank, and my serial number."

My jaw is stiff but is not terribly uncomfortable. I feel very fortunate, but by no means do I feel a match for this man. I know you should not volunteer information. One should never try volleying wits with a trained interrogator. Best soldiers seem to be able to keep cool and very patient.

I am lucky my hands are wrapped and under covers, and my head is essentially totally wrapped. The German asking the questions speaks from notes. I catch sounds of paper. I am soon sure he has talked with the Sergeant who led the search and capture of me, when he says, "Mister Childs."

"Mister Childs?" I think, "What is he going for?"

"You have no identification disks. You are a spy. You will be required to comply with my questioning."

Before response, I take a little time to think. It is obvious he knows I volunteered on capture, "My dog tags are gone."

They either burned through the string, or it broke when the chute jerked me and my jaw extraordinarily hard. I was traveling three hundred miles per hour faster than one normally would pull a parachute rip cord. It seems he has gotten me into a situation where name, rank, and serial number needs backup evidence rather than repeating the same response by rote. I know I am getting into deeper and faster current so to speak.

In my cocoon, I sum up facts. I know Metz is soon to fall and hope General Patton will be along soon. I respond, "I gave my captor my name, rank, and serial number. I was wearing the uniform of a United States Officer with the insignia of an Officer and Pilots' Wings."

He responds, "You are liable to be shot as a spy. This will be continued."

Next day, a seemingly different interrogator, I sense it is the same one, presents himself as a kind person, friendly. He wants to know the name and location of my outfit so they can drop a message saying 'you are alright'. I say, "I am not permitted to give that information."

He responds, saying, "Come now, you want your family to know you are alright, don't you?"

I had a night and part of a day to gather my senses, to think about the situation of my bailout. I determined something had gone wrong. My wing man probably saw the crash but not the chute. I was hit near the target. The plane blew up against a wooded ridge not far, maybe two miles from my target, but less than a half mile in front of me. I guessed I was out and in the open chute five seconds.

I figured the Jug hit within a half mile of me and my parachute. My last words were, "Cover me. I'm going to get the other locomotive."

I answered the interrogator's question because the summation told me a white chute opening over snow at tree top level wouldn't be seen, especially when the Thunderbolt was exploding on a hillside close by at the same time the chute opened. I knew something was amiss, also that the story, if known, would probably be, I was lost in the Jug's crash. I wanted Peggy and Dad to know I survived. I said, "Notify the Bureau of Missing Persons, Washington D.C. if you will, please."

The interrogation quickly changed with another question. He called me Lieutenant, said, "Would you like to talk to the pilot who shot you down?"

I am really glad my face and eyes are not exposed.

He continued, "The Focke-Wulf 190 pilot who shot you down is here in the hospital. He is willing to talk if you would like."

My inclination was to say, "Oh no, it was ground fire." What a clever ruse to get a fighter pilot to start talking.

I had my senses about me in the combat and again at this moment. If there had been any enemy airplanes in the area, this squadron sweep would have picked them up. Fourteen pairs of eyes searching the area for targets would have pounced on an enemy quicker than one could say Focke-Wulf 190. We were used to spotting enemy aircraft, looking for them competitively and continually.

The interrogator didn't know I had seen anti-aircraft guns at about twenty feet, below my wing tip. I figured he was baiting me to say, "There was no 190. I was shot down by ground fire." I'm sure my eyes would have signaled that message. We had been taught their first goal is to start you talking.

I am sure they know my airplane had an 18 cylinder radial single engine, eight wing guns, one four blade prop, also a single seat, armored in back. They probably were digging out both my story and airplane wreckage yesterday.

Our method of armed reconnaissance and strategy had changed since the run across France. Instead of four ship flights attacking, we used two two-ship element attacks. The main formation moved on down a road or railroad while two-ship elements attacked and rejoined like handing off the baton to the next element. This rotation worked well, because dug-in fixed enemy railroad protectors kept their heads down on first strafing passes then threw off camouflage and covers only if no poised Thunderbolts were overhead spotting for tell-tale muzzle blasts giving away gun positions. We won the air war but not enough to ignore enemy fighters.

I chew on the interrogator's question; leave it unanswered. He also leaves the question. I think it is tough to quiz a faceless subject. I expect he catches the night train back to Frankfurt, the center for interrogation.

Chapter 5:
Anne and Louise

Next day, about my fourth day for downed time, I sense the two nurses, *Schwester* Anne and *Schwester* Louise, are standing near my head. They chuckle to each other, then to me. I gather they have a questionnaire for the hospital's information. They giggle because one is jokingly imitating the demeanor of the interrogator. They ask my name, its spelling, and my age. I say my name and my age, twenty-one. One nurse repeats, *"Eins und dreisig."*

I hold up a bandaged hand, saying, *"Nein."* (To myself, I repeat cousin Ed's, *"Eins, zwei, drei.)"*

"Not *eins und dreisig,* but *eins und zwanzig."*

Wow! Things change. One says, *"Eins und zwanzig und Oberleutnant? Nein!"*

"Ja. Twenty-one. *Eins und zwanzig."*

It seems like a sudden lifting of a weight. They have been good to me, feeding and caring, but now they become friends. I have shed ten years. Instead of an aged American fighter pilot, I am a kid near their own age. Thus ends my interrogations. Anne and Louise seem relieved. I am too.

No doubt there was a dossier on me or my outfit at Dulag Luft, Interrogation Center near Frankfurt. In Normandy, radio Axis Sal announced that our clock in the ops tent was off a minute. Probably a good guess. Didn't say which way, fast or slow.

I have been down at least four days and know I have somewhat of a problem; one I present to another nurse who checks on me the last half of the day. I begin to feel the need for a bowel movement. I am constipated with no idea how to get the problem across. My introductory remark works. We communicate. This energetic and positive nurse responds loudly with several phrases resembling shocking news ending something like, *"Mein Gott en Himmel! Nein ziehen! Eins, zwei, drei, vier tag!"*

Her voice resounds through the hallways and bounces from the back walls of the wards. As fast as I had parachuted, I am yanked from bed, pajama pants downed. Creaking wheels signal a rolling commode. I am plopped down on the seat, a greased nozzle immediately administered. Things happen so fast--me? I am just a blur, a minor sizzle on the care meter. Too bad they cannot see my face after conclusion because they would see an appreciative contented countenance.

In the evening, a violinist's playing is interrupted by much activity. A commotion is here! Suddenly I am aware. People move quickly, voices instruct, some sooth. Children cry; a woman sobs. These sounds of distress come from the hall and the ward leading off from the foot of my bed. Eventually a man stands near my head speaking in an accusing voice. His tone also seems to ask, "Why?"

Civilians have been strafed, shot up, in a passenger car on a passenger train. I listen. The man directs anger at me. He speaks in German. He indicates they are victims of something I believe is not common in this area. He laments and scolds for several minutes. I have empathy as a human. It is not an arguable situation.

I am not entirely uniformed. We were given news briefings almost daily in the squadron. I followed the war maps in the ops room at A-79. We have watched hours of movies showing German attacks, sieges, and war in pre-flight training films at Santa Ana. Our evading pilots have come back through the squadron before returning to the USA.

I think how can I respond? My accuser probably has no idea what the Axis Powers have committed during the nearly three years since we entered the war. He is figuratively, if not actually, shaking his finger or maybe his fist in my face. Neither of us understands the other's language, however we might get understanding. I will give it a try.

I want to answer. Nothing I can think of seems to fit. Counter accusation won't work. The hurrying, the moaning, the pain, and empathy are all demanding an answer. In our P-47 squadron, when I was downed, we were destroying only the locomotive on passenger trains, unless requested by ground controller in response to frontline intelligence.

My bandaged head is noncommittal, but my expressed thought may say both our views. My comment comes slow, sincere, "<u>War is Hell</u>."

The man doesn't continue. He leaves. My answer is not original. Simple truths do have multiple origins. I recently found General Sherman said it. So have many others.

Next day, a young man on crutches comes to my bed. He speaks great American, says his name is Ron. He was in the ground attack on Metz, had been shot in the legs and survived. He was hit three days before I was downed. He says this hospital building is a schoolhouse made over. Most of the patients are German Military. He tells me an old man and a boy visit evenings. The man plays the violin; the boy slips Ron an apple. The boy takes the core away when they depart. (Two months later, we would have eaten the apple core.) Ron says we are on the second floor and the surgery is on the floor below. (Later, we are ward mates in a POW camp for a week or so.)

Likely, what I thought occurred over the next three nights of dreams all happened in the first twenty-four hours. Three times, I dreamed that Patton's Forces were coming into town and I would be liberated. My most fantastic nightmare ended with a big banquet table with me the roasted centerpiece, apple and all. The scent of an offered apple may have triggered my nightmare.

Ron Bleecker visits me twice in the Kaiserslautern Hospital ward. He helps me to orient as to location and time and the daily routine. He is assuring help to me. Through him, I am able to figure out days and dates and to make a few notes when we meet later.

Soldiers singing in the wards are not too different than cadets singing. Some songs are of universal familiarity. I thought of we kids singing around the piano at Olex Grade School and my Dad's pleasing baritone in Sunday School at our country church. I became Sunday school secretary. C I C was my class for high school students; stood for "Class in the Corner." I had worked my way up from Sunbeam then Sunshine, through Four Leaf Clover classes. The little one room church was midway between the two room schoolhouse and the cemetery. I graduated with the biggest eighth grade class in years, six students. A girl was top student. I was second.

Our class was twice as big in high school, seventeen miles north on the Columbia River. She was valedictorian. I was salutatorian. She read literature. I read "Flying Aces" magazines and "The Flying Carpet."

Around the school piano, we sang songs from the "Golden Songbook." I like singing. Both the Germans and we on the other side of the lines sing "Lili Marlene."

A voice from the hall bed behind mine, comes through my bandaged ears. *"Herr Leutnant, you singen 'Santa Lucia'?"*

I grin in my cocoon. I know "Santa Lucia;" I answer, "*Ja,*" but decline. I like his invitation.

About the sixth day of my hospital stay, Nurse Anne and Nurse Louise come around after normal ward work is completed. They begin taking away my head bandage, removing the wrapping down to what seems to be pads over my eyes. I have not worried about sight because I could see a little after the fire. The German doctor takes over, removing the pads which seem thick and messy. He asks for an instrument which feels like a small metal spatula. He clears my lids, separating the mass over my eyes. The doctor's face is very close to mine. The gunk is partially cleared, when he asks, "*Herr Leutenant*, can you see me?"

All I see is a monocle with an eye in the center.

I reply "*Ja, Herr* Doctor."

He replies, "*Gut!* Your eyes I have saved."

I sense a bit of arrogance in his tone, but am happy to see even though it is cloudy. I am not particularly disturbed or elated. It is "wait and see time." My bed is in a raised position. I sense the nurses close by. I lift my shoulders slightly, looking carefully. I see two faces leaning over the bedside. I say, "*Anne*?" She greets with her cheering smile.

I see, close beyond, over Anne's shoulder, a second cheering countenance. I call out again, "*Louise*?" She beams.

Their charming smiles are healing therapy to me. It is great to make contact with these kind young women who have nursed, fed, and cheered me.

My bed remains in its place in the hall by the nurses' station on second floor. The entrance seems to be around the corner behind my head. Eventually I count visitors in and out as a pastime. About two days before I ship out, a singing group comes caroling in the evening. They sing in the ward behind me and the ward off the foot of my bed, and continue on through the hospital to additional wards on the floor above.

They are faintly audible as they proceed. I think of great Christmases in my life, especially the two with Peggy. The singing is a cappella with many songs in common with ones we sang at home including "Silent Night." I am aware that the singing will be ending soon. I am about as blue as I have been. I hear silence, maybe snooze. Have they left? Suddenly, I catch the sound of many feet close by. They are done, on the way out. I want to say thanks. Then--not a sound.

Suddenly, from wall to wall, halfway around my bed, maybe two or three deep, beautiful voices burst into a magnificent choral number. Their kind gesture became a very personal gift; one of hope for both them and us. I say, "*Danke Schon*." "Many thanks."

This is near the end of my stay in Kaiserslautern. I came after dark November 17th. I leave after dark November 26th. I arrived on a stretcher and leave on a stretcher.

The ten day interlude, attended completely by German nurses, one of whom spoke a little English, stays with me lifelong. I saw only the German doctor's monocled eye and the lovely faces of Louise and Anne. I hope each married and have children; someone I can tell how their wonderful mothers improved an American soldier's life.

I never got hungry while at Kaiserslautern Hospital. The routine medical did not seem to vary. I was fed with a spoon from a bowl. I do not remember any bandages being changed other than the one time for my eyes.

I know I have severe burns all the way around my neck, and under my chin, backs of all eight fingers, nails, my forehead, around both wrists, my right thigh, and the backs of both legs. My jaw is stiff, thumb nails smashed, my knees untested. I drank lots of liquids. My temperature and pulse were monitored; that is about it.

My clearest memories are kind gestures from the nurses, the civilian singers, German soldier, and from the American soldier, Ron Bleecker.

I have said I was visited by Ron Bleecker only while in the *Krankenhaus* at Kaiserslautern. He helped me much with his orientation talk. I was impressed with this youngster of about eighteen. He was on patrol through the Maginot Line at Metz when he was shot through the legs. He lay out in the field a couple of hours, jumped up running and was hit again by rifle fire. He made it into one of the sea shell-like isolated protection shelters and remained there until dark when he was captured.

Ron came by hospital train, arriving at the hospital November 16th. I arrived on the 17th. We both left on the 26th but didn't see each other again until December 4th in Stalag IVB. Our contact continued when he came into our ward. He referred to it as the Convalescent Hospital.

Ron's story is a part of the Kaiserslautern story. Ron was a bright high school graduate, seventeen years old, in the middle of the war. His father was an engineer for Curtis Propellers in New Jersey. Ron volunteered for a special Air force flight line pre-cadet program and was swept into the infantry and 3rd Army for the drive across France. After the war, he and I each tried to contact the other. Finally, fifty-three years later, his daughter found me on the "people finder." Later yet, Ron found my 1945 letter and picture in his Dad's files.

I have talked about the good treatment I received from the German staff. Ron, however, experienced much more than I. Briefly, his wounds were not tended by the doctors. They raised the sheet, looked at his smelly leg, and dropped the sheet muttering, "*Kaputt*."

Ron knew a little German language. In our conversation at Stalag IVB, Ron told me how after the doctors left him to die, two nurses became indignant. He remembers a third nurse, Marie. She and one other loaded him onto a stretcher and onto a gurney, got him down the stairs to the surgery, and more or less like a battering ram, pushed their way into the surgery, shaming the doctors into caring for him. Ron described a female doctor as fiendish and frightening. She picked up a knife similar to a carving knife, touched it to her thumb, then pantomimed the cutting of her throat.

Ron told me the story without fanfare, plain and simple. He felt another few hours of fever and his limb would have been too late to save. He also perceived the good side of staff and folks same as I. His story is striking with several remarkable events, made less remarkable when you become acquainted with him and his family. His POW experience finalized in an escape, enabled by Ron and three companions.

I've mentioned his apple violin story. In recent years, Ron showed me the first letter received by his family, April 4th, 1945. Mailed in Germany, it was written by him on November 22nd, 1944 and bore the cancellation mark 11.22.44. The letter has a censor seal and swastika. Ron gives Schwester Marie credit for saving his leg and for mailing his letter home. Our sparse notes tell us we both left the hospital on November 26th arriving at Stalag IV-B November 28th. Our POW serial numbers are in sequence at IV-B.

On the Hospital Train, it is warm; my bunk is comfortable. Several American wounded are on the car. We leave Kaiserslautern in the evening, travel most of the night. We are given bread and cheese and something like coffee. I am helped and treated similarly to hospital care. Ron is not on my car.

In the morning, several American GIs come to my bunk, saying, "We are parked in a narrow canyon, heavily wooded on both sides."

They are concerned about a train of flat cars loaded with a Panzer unit parked alongside our train which has its Red Cross hospital markings. I shudder along with the others. My conclusion is if one of our squadrons should spot the Red Cross markings, they will knock out the engine of the military train and give the hospital train a short time to move away, then clobber the armored unit, tanks and all. I am relieved to learn the weather is very bad outside.

We don't get hit by aircraft, but we spend a little time worrying. There seems to be a lot of railroad traffic, and we aren't a priority. We spend all day and night until early next morning getting to Leipzig. An American interprets, saying, "You will be leaving the train at Leipzig. You will be put on another train. It will take about an hour to reach the prison camp."

I was dressed before leaving the hospital in my dark green shirt. My insignia has been removed. Other clothing adds up to my T-shirt and boxer shorts. I wear a considerable amount of bandage. My field boots are tied to my stretcher. I am between folds of a wool blanket over canvas.

The train stops. I am steeply handed down to the snow covered ground then carried across railroad tracks to a freight train. It is full daylight, very cold, especially coming from the 'luxury' of the warm heated German hospital car. I am lifted up into a box car, like many I had strafed with eight 50s. The door is rolled open a little more to get me loaded. We did a lot of work on German railroads with Thunderbolts. I hope I am ahead on that score but don't seem too likely to continue my winning streak.

My stretcher is placed inside but across the door opening. People whom I think are German guards stand around a crackling fire, maybe in an empty oil drum. We used one for ice skating at home. I think, "an hour," I can handle that, no sweat. I am right about the "no sweat." The car door is left partially open, even after the train starts. I feel the cold. My left hip with no bandage is exposed to outside breeze with only a little protection furnished by one layer of blanket. The sun's rays and the breeze are about a standoff. Cold!

I feel the bottom of a soldier's greatcoat touching my bandaged head. It brushes over me often enough to agitate my chill. The stretcher has short legs and a canvas bottom. Mine gets colder by the mile. This is not exactly inhumane punishment. It is more like disdain. I do not say a word. I think I will ride it out. Certainly would not take heat away from anyone to move the stretcher close enough to get benefit from the fire.

I have a pretty good acquaintance with freight cars and steam locomotives in our county. A daily train chugged the thirty-eight miles, Arlington to Condon, six days a week. I could hear the mainline trains on both sides of the Columbia River from our high school study hall, sometimes even from a goose pit, ten miles south at our ranch. I ran a cross-country route along the track south of town daily in the spring, four to seven miles. I helped unload slab wood from a sawmill rail car on a siding about five miles north of our ranch. I ricked wood outside after the woodshed was full. We had a two years' supply.

A couple weeks after the wood was ricked, I went to 4-H Club Summer School. A skilled hatchet throwing chaperone set up a huge slab of wood for a backdrop to his target. He was so accurate, I was really impressed. The first thing after coming home, I systematically hatcheted about three ricks of wood. The car held about twenty cords. We burned split slab in the kitchen cook stove.

I attempt to determine the direction we are traveling by time of day, warmth of sun, and starting place. I figure I might be one of the first of our outfit to tour east of the Rhine by rail.

By the time we stop and I am handed down to what sounds like French speaking people and placed on the surface with bottom of stretcher reaching into deep snow, I am as cold as I have ever been, including my second grade experience, falling through the ice under the bridge at the Olex Store. Mrs. Campbell kindly got me over to their house, peeled my wet clothes off and gave me some dry ones. That was not in the same league with this cold.

It chills me a lot, knowing people will deliberately make life miserable for a person. So, I am pleasantly surprised when a greatcoat is thrown over me. When a second one is placed on top, I am warmed from the inside out. After two more coats land on me, I want to put one between me and the canvas bottom. My bearers speak French. I begin to thaw. They walk at a steady pace. With sun shining down, carrying a stretcher is warm work. Their coats are my good fortune. It seems a forty minute walk. They set me down on the trail once. The gradual thawing process is a blessing.

Chapter 6:
Gold Stock to Wedding Band

Roll call is outside the double doors to the Portland Union Railroad Station, February 23, 1943. Our goodbye is both tough and a beginning. Peggy and I are in love and close. She gets in the car. I pick up my small Samsonite and join lots of young men arriving and waiting.

Two or three are friends from college. I recognize a couple names out of the past when roll is called. This is not sophomore ROTC. We are civilian recruits to the Sergeant who calls names from a clip board's pages. He stands on a baggage cart. "Childs," sounds almost like two syllables and my first initial, "A (initial only) David" fits me.

I answer, "Here," the first military word I have spoken since swearing in nearly ten months ago.

We board several chair cars and hook onto more. We come from all directions and now are heading south. We are volunteers, looking forward to aviation cadet training.

The landscape is familiar along the banks of the Willamette River. The train to 4-H Summer School traveled this route stopping in Albany, then we bussed to the OAC campus, Corvallis (later OSC, then OSU).

The 4-H Summer School train went on the west side of the river just one time, through Newberg. The track crossed Peg's folks' farm. That was close but not as romantic in our before college meeting than our go-between relationship with Mount Hood. I looked seventy-five miles west to "my" mountain; she looked seventy-five miles east to "her" same magnificent mountain.

This ride and not knowing to where, leads me into reflection and brings to mind, times and places on both routes along the river. Dad, mother, aunts, uncles, cousins, all people dear to me; Oregon City, New Era, Salem and Albany. Now in less than a year it is she, Peggy, and our wonderful life, that makes my lashes misty. Peggy tells me in a letter,

"When I came in the door at home, little sister, Mary Ann, now almost fourteen, is at the piano, pounding out and singing, *"This Is the Army, Mr. Jones, No private rooms or telephones"*...

Peg added, "I pounded her."

Many of us signed from colleges with ROTC training. I had four terms of weekly drill. Our sergeant at BTC # 8 Fresno, California sorts out a platoon of us with previous ROTC training and thoroughly works us through close order drill. I overhear the Drill Sergeant chortling to the Lieutenant, "Sir, those men got their uniforms this morning. I've got them standing retreat this evening."

Looking back, I know my father looked ahead. He deserved a boost; he made it go far. First year (1928), we had a very fine wheat crop and a great price. Dad bought equipment, painted and wired the buildings; hung bare light globes in all rooms in the house. I heard him tell people, "Cavy will select the light fixtures when she gets home."

He had the back porch roofed and screened in. He also enclosed the cellar with a new laundry room and attached all to the kitchen as part of the house.

Dad invested in Canadian Gold mining stock. It was a significant benefit for us. It compared favorably to his buying three Shorthorn weaned calves, registered purebloods from Northwood Farms, northwest Washington. They arrived in wooden crates at Arlington depot; two heifers, Bell and Queen, and a bull, Louis.

During the Depression, Dad received twenty to thirty dollars cash dividend every month from the gold stock. He paid the expenses of his 4-H Club Judging team and supported the school's playground equipment those years. His laying hens paid for our groceries. Registered bull sales became important. Our small cow herd increased. Wheat crops were poor and so was the price, however almost broke even, if you didn't count labor.

Dad planted a plot of Crested Wheat grass for seed. His encouragement came from Russell McKennon and E. R. Jackman, both gifted communicators from "the College" Extension Service. They and the drought-depression hit Eastern Oregon's grain and grass lands with a full swing about the same years. McKennon had played second base.

Jackman had a hitter's eye. Both showed horse sense.

Crested Wheat grass seed at one dollar a pound sounded outlandish. Dad built a light box from roofing metal he salvaged from the rubble of a burned store. He attached the box to the back of the sickle bar on our ground-powered two-horse mowing machine. The precious seed bearing hay was pulled back into the box by Cousin Ed's visiting sister, Mary Louise. She saved the day. Sis was in high school. She raked the dry headings into the box with a bamboo leaf rake. Dad drove the mowing team. I drove the 1928 Chevrolet one-ton truck with stock racks. Ed forked the headings onto the truck.

Ed and I flailed out more than one hundred pounds of seed on the wooden floor in the feed barn. We cleaned it through a small hand-cranked Emerson Kicker grain separator; netted one hundred pounds in two fifty pound sacks.

Ed was a year younger than I. He brought big city Portland kid knowledge, and we shared open country ranch life. We had a wonderful time for a couple months for several summers. We herded cattle, hunted rabbits, and swam in Rock Creek and the John Day and Columbia rivers.

We did ranch chores; even teamed to pick up and haul the seed and feed wheat, fifteen sacks a load, from the field into the feed barn with our 1936 International pick-up. We kicked and passed a football and high jumped, long jumped and pitched horse shoes. The part I liked best was hauling straw with the team and header box. We only got about three loads a day.

I listened while he described all the movies he saw. We sang all verses of "Home on the Range," "Strawberry Roan," and a few of "Buster Jinks and Sage Brush Sam." Ed brought "Sixteen Men on a Dead Man's Chest," "Abdul Abulbul Amir" and I added "Little Joe, the Wrangler." Ed raised with "The Daring Young Man on the Flying Trapeze."

We slept on army cots under stars and locust trees in the backyard the years before we became the crew and moved into the ancient bunkhouse. One of the many inscriptions on the bunkhouse wall follows-

> A hobo sat on a box car
> and his feet drug the ground--Longfellow

I kept my savings account growing; Dad started it when I was one year old. He told me this when I went with him to the head office of the US National Bank in Portland while we were down for the Pacific International Livestock Show. Dad was leader of the Olex 4-H Beef Club. He took the Livestock Judging team to participate in the judging contest. I was nine but not on the team. October was a great time, World Series games were broadcast on radio throughout the exhibit and show barns. Political party booths dominated choice locations in the main pavilion, handing out buttons and literature for presidential candidates and congress and State Government. In those years, 1932 through 1940, baseball, politics, and people attending the daily horseshow and rodeo were what Dad would call educational.

We went to the bank first time down, then again two years later just to get my savings account pass book brought up to date. Each year it would get a bump from the sale of 4-H animals. We mailed in deposits. The last was in my freshman year at Oregon State College. This steer came from our own Shorthorns. He placed near the top in the 4-H market steer class. I sold him in the 4-H auction.

My checking account started before I was fourteen. The first check was to Montgomery Ward for a bolt action 20 gage shotgun and a slide action twenty-two rifle. My 20 gage double barrel with hammers was becoming unsafe and my ancient twenty-two was worn out. The check was for less than twenty dollars.

Dad gave me responsibility with a loose rein and his example of right and trust. He taught me to drive the ranch equipment and straw hauling team and let me drive his car to school at fourteen when I passed the driver's exam for a school permit. He set rules. His method of training pups was to have them drag a smooth cotton leash long enough to step on if the voice control was ignored. I mostly listened, got the message.

He helped me get a loan from the bank at seventeen. I bought twenty head of feeder steers and fed them to market weight. He pinch hit feeding for me on days I was away for games and track meets. The profit from the steers and a couple of scholarship awards were enough to get me through the first year of college, 1941-42. The feeder steers were contracted from Scott Brown. We signed the contract in the lobby of

Hotel Condon. Scott and his sister, Linnie, trailed two-hundred head to Gilliam County from Burns, through two mountain ranges and a lot of road and trail; he said 200 miles. Steers were "dropped off" at other ranches, some for Scott at their Rock Creek ranch near Devil's Gap. The last twenty were delivered into our corral on Shutler flat. Mrs. Brown tended camp. I was impressed with their cattle drive. My feeding enterprise succeeded.

My lawyer uncle wrote to me saying, "There are funds for you to attend any school."He recommended his schools, Oregon State and Michigan. I asked my father if this was acceptable. Dad said, "Yes, because it is from money owed, borrowed from your mother's account." I was already enrolled in Oregon State with a room in Weatherford hall.

It is welcome and with scholarships paying tuition, I will not have to dig into my savings. My uncle put no strings on it. He deposited three checks to my account, over two years, in a Corvallis bank.

Summer of 1941 is a land mark for me. The previous three summers I had worked on a large neighboring ranch. I started as header tender and did some of about every job on the crew. I also did Saturday work ranging from driving a Caterpillar pulling a wide hitch of weeders to doing "summer-fallow polo," hoeing scattered weeds from horseback, with a long handle hoe. It was great training. They gave each of their combine harvesters a single digit number. We, not to be out done, painted "IT" using two letters on our one machine.

I enrolled early in mechanical engineering. I wanted to qualify for one of the military academies and earn an appointment. My senior paper in high school English was entitled, Army and Naval Aviation in the United States. The six thousand words featured both schools and the training programs and requirements for both services.

My Portland cousin Ed, Dad, and I pioneered a fine little two-man harvest outfit in 1941. Ed and I had a great run with only one serious interruption. I tell it here because it may have influenced my response in Series of Events.

Ed, smarter than me, flunked in the sack sewing position planned for him. I was dethroned from tractor driver to become king of the combine's dog house, as combination sack jig and sack sewer. The wartime shipyards in Portland would keep Ed home the summer of 1942, so this is his and my valedictory year.

The crop is standing up good so Ed who also raised and lowered the header didn't have too busy a time. We are traveling west, with hardly enough breeze to move the dust away and behind us. A kernel of wheat wedged in the knife groove in my sack needle. I pull the bell rope signaling stop. It is time for mid-afternoon greasing anyhow. I cut the twine with my pocket knife and clear the needle. I stand looking out east across the highway. Suddenly, lazy day leaps to an end, smoke is curling up behind one of my neighbor's combines. Fire is moving into standing wheat. Ed and I jump to action. He helps unhitch and relinquishes "his" tractor. I have three years of spreading cricket poison-bait and mowing wheat field right-of-ways with "my" pride and joy; it travels sixteen miles an hour in road gear.

In three minutes, I wheel over to Ike's shop and holler, *"Fire."* He directs me to a three bottom plow and hooks it to the drawbar with a chain and clevis and wrenches the nut on tight. This hook up is unwieldy and swings to and fro traveling across the field until I trip the plow into the ground, close to the spreading fire. I do pretty good putting out fire until straw plugs my plow and stalls my tractor to spinning wheels. The breeze picks up at the same time. Fire is engulfing my straw loaded plow and coming at my tractor. Only one way the tractor can be saved. I hit the clutch and shift into reverse to gain slack and running room, about seven feet. When the draw bar backs into the burning mess, I shift into 4th, fastest power gear. I give her full throttle and pray the chain will break. If it does, we will be out free. If it doesn't break, I sail over the hood and land on my feet running and the stalled tractor burns. The loud snap of breaking chain link is beautiful. Ike sees the caper and meets me with a foot-burner single bottom plow in his model A Ford whoopie pickup. He pins the plow hitch directly to the drawbar and grabs its handles. He soon tuckers out and discovers the plow stands upright pinned solid to the draw bar and throws soil, smothering fire like all get out. Men follow with wet sacks and shovels squelching spot fires.

In May, 1942, I wrote asking Dad for a letter of permission to join the Army Air Force Cadet Program. A few days later, I wrote, *"I won the mile race in the freshman track meet with the University of Oregon,"* also to thank him for his permission letter for the Army Air Force: I added, *"I gave Peggy my Sigma Chi pin…just good friends."*

My ten year-old hired "hand," neighbor boy, Bill, and I did the wheat harvest. His and Ed's similar willing, "it can be done" attitudes made another harvest fun. Bill drove tractor. I operated the combine, and sewed sacks. We hauled the sacks of wheat to the warehouse in Arlington after harvest. I lifted them onto the truck bed and Bill rolled them into place; they weighed 135-140 lbs each.

Peg wrote about once a week. Dad would stop along the road when I was doing interruptible work. I knew I had a letter from Peg and would jog over. I visited her at her folks' farm the 4th of July. We picnicked along the banks of the Willamette River flowing by their farm. Her thirteen year old little sister, Mary Ann, came with us.

Dad was in the hospital after harvest. Two of my fraternity brothers were in a tragic auto accident. They were twins; one died. The survivor was in the same hospital as Dad. The concluding services were in the Willamette Valley not far from Peg's home. I visited her before heading home.

Dad was soon home from the hospital. He probably knew a couple months sooner than I did, his son might be writing home more than just his ordinary letters. In November, I wrote for parental permission to get a marriage license!

I was to run a cross country race at Spokane on Thanksgiving Day. As soon as I got back, Peg and I were going to Eastern Oregon to see Dad. I had a cold and fever on the trip to Spokane, but thought I'd sweat it out in the upper berth on the Pullman. Next morning, I felt weak but OK. No fever. I started well and ran respectably the first two miles. The race was four miles on a frosted golf course, about 28 degrees temperature. I suddenly ran out of gas. I had never started a race I hadn't finished and never lower than the top three since I placed seventh in the State of Oregon High School track meet as a sophomore.

I felt pretty good back at Corvallis, but stopped by the infirmary on the way to pick up my Model A at the garage. The infirmary was handy and I had a slightly stiff neck. The doctor gave me some aspirin. I thanked him and headed for the door. My hand was on the knob when he called me back and stuck a thermometer in my mouth. He put me in the hospital. I had pneumonia. Peg and I wouldn't be arriving at the ranch in six or seven hours, and Dad would meet his future daughter-in-law when she met her father-in-law.

Peg and I were married Friday, a week before Christmas, December 18, 1942. We went to the coast for our wedding trip; wartime, black-out-curtained windows; no lights, no people. Yeah!!

Dad would see us after Christmas when we come to the ranch. Gas rationing and livestock care kept him home.

Our plans were for me to request active duty, to not return to school winter term but for us to stay with Dad on the ranch until my call up.

We drive to Arlington the day after Christmas. The Columbia River gorge is spectacular anytime. There was a skiff of snow; all black and white. The gorge and trees give way to bare basalt bluffs and bunch grass.

We don't stop in Arlington, drive south on Highway 19. At about twelve miles, the highway breaks out of the canyon onto windswept Shutler Flat. I point to our landmark, "the tree." Our land corners here and follows the highway. I point out the tops of our buildings a mile to the west. We turn right onto the first road, going west and follow the section lines for a mile west and a mile north. Could be dust or mud. Vehicles are visible a mile before they arrive.

Dad meets us at the front gate. It is cold. He is wearing a jacket and stocking cap. Dad is cordial and Peg is herself. Both are great! She is twenty. I'll be twenty in a week. LW likes the little Willamette Valley girl. She captures his heart, too.

Dad has roast beef surrounded by potatoes, carrots, and onions ready to come from the oven. Smells wonderful! The house is warm and neat. He wears bib overalls and a blue shirt. The breeze has a bite. It is colder on the flat than down on the river. My upstairs bedroom is just as I had left it in September. We added a big quilted comforter.

My eighty year old neighbor, Jud Baker, a jockey who had ridden the Mid-Columbia Circuit and gone to the Kentucky Derby one year "just to see it," loans us one of two horses he keeps, also a saddle. He rides Bud, his big red retired race horse, two miles to Dad's to pick up his mail. He insists Peggy have a horse to ride.

We settled in living about like winter weekends at home; Hunted ducks, rode horses down to visit cousins Frank and Daisy. They were fun. We hauled loads of gravel to the front of the horse barn space, hand shoveled on and off; also transplanted some good sized trees from Dad's

Gold Stock to Wedding Band 41

plantings; played three handed pinochle.

New Year's Eve, we went to the banquet and dance. Another time, at a Grange potluck, Peggy took a simple Jell-o salad in a green glass bowl. She later smiled saying, "Your father was proud of me."

Peggy to her mother, Margaret: excerpt

Dave's father told me if I wanted to call him something, to call him Lawrence. He is awfully nice, and likes me. He and Dave tickle me, they have been trying to do things to please me. The other night Dave got his father to take down seven calendars in the living room and his father got him to take down all his track numbers. They had both mentioned to me privately what they were going to have the other one do. (The decor didn't bother me-pc)

We ice skated on the alkali flat ponds along the railroad track. We bought Peggy white shoe skates. We fed cattle, read, listened to the radio. We were the only ranch house with power line electricity. We visited Peg's folks Valentine's Day, a week before I was called up. We brought back an Australian Shepherd pup and named her Flicka. After we left, Dad and Flicka strongly nicked. She was an extraordinary participant in the lives of all three of us.

Time sped by. We told Dad goodbye in the bus depot in Arlington on February 19th, 1943. We'd taken Juddy's horse to him the evening before and rode my horse, Lightning, double coming home.

My uncle's College fund boost probably was not meant to include a honeymoon. Heck, I hadn't either. Our "looked paid for" rings and wedding trip came from my East of Mount Hood bank account. The education fund's overrun stays in the Corvallis Bank until we return.

May 1942 through February 1943, our lives had moved into wartime awareness. I added a survey course in aeronautics fall term. Peg added thoughts of marriage and leaving home to her school schedule. The war was preeminent to our plans. We would do our share. We were at the start and on track.

Chapter 7:
Tracks for Sure

Basic Training Camp number 8, Fresno, California shuffled us well, more like a mixing machine for West Coast recruits, red, black, brown, white, and then us, for *aviation cadets.* It is good. We are exposed to everyone's army everywhere, on the drill grounds, the obstacle course, at the PX and the theater. Haircuts evened us; shots floored a few; calisthenics pained our limbs. Moving stuff forth, then back twice again, affront my farm-kid sense of reason; everyone else's too, until later, we sort it out as to why, now, hanging our coat hangers all one way, and doing precision corners on our bunks may save some necks, our own. Altogether, it makes us coachable. There are rumors too. Only one was true, yep! Mumps! Our barracks gets quarantine for three weeks, then we are ordered on our way. Just like livestock; we don't know where, we just go until we are there.

The second train ride introduced a new subject. Blackjack! It was like farming. You guessed a lot but had to observe and reason also. In the end the odds won.

We went north beyond Portland, over the Cascades to Ellensburg, Washington, Central Washington College of Education. It was only one-hundred miles north of home. Peggy visited five weekends; Dad visited once.

The obstacle course was perfect for me. I vaulted fences between cattle and horse pens when doing feeding chores at home. This course used a lot of preexisting structures and some handy landscape. Start was down the street, straight at and over a board gate, run across the rodeo arena, over the arena fence, and up the stairs to the top of the grandstand built on the face of a steep hill, hop over the top seats onto the ground, turn left, run along the ridge, down steep bank, cross over the creek on a plank. Cruise around to cross a bridge, then jazz it up heading back to start.

Tracks for Sure

The accelerated classes became even more interesting because of our attractive math teacher. Our section was assigned spherical trigonometry. Even though she was eight or so years senior to most of us, we considered her "all right." She dressed neatly and certainly was not unpleasant to see.

Afternoons, our section flew in 60 hp J-3 Cubs. We got ten hours dual with no solo. We were required to sketch and write the series of training maneuvers; turns, stalls, climbs, and spins; everything for each flight. I still have my spiral notebook.

From the Cubs, we saw the spring thaw, with water flowing and felt the Chinook wind. We learned West Point type discipline. The beautiful surroundings made indoctrination, including guard duty, endurable.

One prized vision from the month and a half spent at Ellensburg is the image of the math teacher and her circle. She scribed it perfectly at the speed of light. The circle seemed to appear from nowhere, didn't interrupt her thought or talk. She, not the blackboard, had our attention. Her shoulder was just the right height for a pivot point. Her unnoticed thumb and finger most deftly held the chalk, arms by her side; it began from the bottom of the board and ended in one full circle sweep with the ends of the sphere coming together exactly at its bottom.

Every class, we would get one and sometimes more perfect circles. Seeing the actual creation eluded us. The speeding chalk didn't grab our attention until it screeched the circle's completion. We never saw the start. Few saw her complete the arc. She won most times, and so did we.

Our formation leader, at least twenty-one, sophisticated, third year Engineering, University of Washington, was careful to keep us abreast of all developments. While marching to our next class, leader would, in cadence, ask, "Who saw the circle?"

We, in cadence, came back, "No one did."

We were not asked to detail any of her drawings. We each hoped to not ever have need for her teachings. Not because we didn't learn, but we wanted to be fighter pilots. Our missions would not depend on our trig. Radar and our long count turned out to be 19th TAC's fighter pilots' best friend. They did the trig at speed of light for us. However, it didn't replace the memory of the perfectly proportioned math teacher or the Central Washington scenic setting. I wonder if she knew of our captivation. "Bet she did."

Next: Santa Anna:

Nine weeks compressed into four and a half. Time for physicals and testing, then assignment: Pilot, navigator, bombardier or move on. Classes were Morse code and aircraft recognition; Allies and Axis, also ships of both. We saw hours of movies of the German attack siege on Russian cities.

We had lots of calisthenics daily. I won the mile race at the graduating class track meet. I passed a cadet who ran for the San Francisco Olympic Club. At the end we were very close. I was awarded first. It was my best race. His also.

I paid a dollar for a coke at Coconut Grove, famous for Big Bands and Hollywood stars; went paddle board and sailing at Long Beach. College friends had invited us.

Four of us Santa Anna squadron mates went all the way through training and into a Fighter Squadron together. We became acquainted in Tucson for Primary flight training. Around and beyond Los Angeles, little Primary planes were flying. We saw some on the way out. P-38s trained overhead and over the mountains in sight and sound of Santa Anna. We really were pumped.

Ryan Field was hot and wet and dry in August with a thunderstorm nearly every afternoon. We quit remembering the spherical trig teacher when we got to the flight line and saw those gorgeous low-wing Ryan PT-22 trainers. They landed at 60 mph, had inertia starters, low single wings, and great civilian instructors. Mine, Mr. S, graduated all six of his students. It didn't take him long to get his point across. He was busy. He brought his pretty wife to our graduation party.

Another excellent instructor, after seeing one of his student's low approach to a landing at an auxiliary field, sent him out to collect shreds of tall grass mowed by his prop. The instructor put bold letters on a sign saying, "Next to the last straw." Sent him to walk the flight line with the sign hanging around his neck, straw attached.

From Ryan, we travel east to "out west" to Pecos, Texas for Basic Flight School. Peggy came to Tucson before we graduated from Primary. She and a cadet friend's wife teamed up to drive to Pecos. Peg eventually found a place to live; helped keep house for a flying officer's wife. They had a new baby. Captain White's parents lived at White's City at Carlsbad Caverns. After the war, Peg's parents stopped and visited with them. Their son was lost in the English Channel.

Pecos was still a frontier type town. The butcher shop section of the grocery store seemed like a good place to buy a chicken to fix for Thanksgiving dinner. Peg, quiet sweet farm girl, never even blinked when the man handed her a live hen with feet tied together. She took it home. Captain White chopped off its head. Peg dressed it. She and another cadet wife had a tasty though somewhat tough dinner. The cadets flew on Thanksgiving Day.

When we cadets got off the train at Pecos, four coffins were being loaded aboard the express car; two instructors and their cadets. The BT-13 was a good airplane, great to learn formation and good to fly. It needed positive coordinated control. It could stall and snap to a spin faster than the spherical trig darling spun her circle. The cadets at Pecos sang their version of the "Volga Boatmen." "Watch That Last Turn," (da da Da daa).

Pecos ground school had one instructor whom I thought was one of the sharpest people I had met up with. Captain Blaco taught weather. He possibly doubled as a psychologist. He called on each of us, spread the conversation around. His causticity was clever, just a nick now and then. His skill with language and kids was like a slalom skier carving close by bamboo sticks. He talked with us for thirty-five minutes and taught us weather in fifteen, each class. He was good. We learned.

Captain Blaco had us in October and November, 1943. Nearly two years later in September, 1945, I was walking through a long hospital corridor in Van Nuys, California. I glimpsed him going the other direction. I turned saying, "Captain Blaco?"

He may have been Major or Colonel, I never looked. He turned quickly, looked like he was computing and pointed with his hand, said, "Childs" as he stepped toward me. He knew me. There was no question in his voice. We chatted a few moments; great man. I had just received orders for transfer to Dibble General Hospital at Menlo Park, California for plastic surgery.

Pecos to Luke Field: Peg teamed with Alice from Brooklyn, N.Y., another cadet friend's wife, for travel partners to Phoenix. The girls found a room for rent in a large home owned by a Jewish family. We traded weekends staying at Polly's or going to hotels.

After graduation, Tom was assigned as an instructor. He was killed in a training accident. Alice said, *"Taking Tommy home on the train was the worst thing I ever did."* She never remarried. After more than sixty years, we still exchange Christmas cards.

Flying AT-6s went well. We had started flying with ten hours in a Cub near Washington's Cascade Mountains and ended with ten hours flying P-40 Warhawks over Arizona's White Tank Mountains. We got time off to order our uniforms before graduation in February.

One of my friends, Dean, from a high school in our league in a neighboring county, and his instructor were killed in a crash at Luke. His name was already on the orders listing graduates. Peg and I became acquainted with his parents after the war. I knew him at 4-H Summer School since we were about twelve years old.

GRADUATION 1944

I was sitting, hands free, along for the ride because my hours had come up a few minutes short for the total training package. I liked to think the mostly desk-flying pilots participating in the graduation ceremony review flight wanted the assurance of a currently rated pilot along on this crowded flight for show. We were in echelons of slightly rippling formations. To us in back seats, it was shaky, however, military and flying discipline took over. Yet, for scared-anxious, this flight nearly equaled my first combat among flak bursts.

Our gold bars shown; trousers were creased and bags packed. Peggy pinned the wings on my blouse. An army photographer snapped our photo. It was sent to The Oregonian.

Eleven of us were presented pilot's wings engraved, *"PT Champs of Luke Field, Class 44B."* The teams were by alphabetical selection and given designated names. Our team, the Lions, would scatter across the U.S.A.; Massachusetts, Pennsylvania, Washington, California, Iowa, Oregon, Texas, Alaska, Michigan, Florida.

Right at four hundred new officers were all heading out to assignments. Most of our fifteen day leaves would be consumed with travel. Peg arranged early for tickets. We lucked out; went by Pullman to

Portland; stayed overnight with Peg's folks; next day by Greyhound up the Columbia River Gorge to Arlington and home for a touch of the past.

It was nice to be with Peggy every day. We had three and a half days at home. We walked along the partially frozen creek. Skeet training at Luke provided a two-duck feast for Dad and us. No guess work.

Luke Field's skeet training officer, Captain Jules Cuenin, was the best. Held his shotgun across his chest, one hand on slide, other on the grip. Doubles, singles, or coming straight at him, two clay pigeons at a time, high or low, he got them. He was also very good at instruction. First he told, then he demonstrated.

My hunt was to try what I had learned. My shooting had been OK, but he made it sure and easy. Formation flying, shooting and learning gunnery, aerial and ground, were things I knew a little about, enough to learn much more. My cadet flight at Gila Bend gunnery school chose me to lead on our qualifying pattern. Both air to air and air to ground were exciting. I qualified expert in both. On the farm, driving Caterpillar tractor and pulling a fifty foot hitch of farm equipment at three mph along a fence or through a gate was similar, maybe tougher, than flying on a wing in tight formation at more than fifty times that speed.

Coming home for Peggy and me was sort of a graduation and a bonus beginning for us together. Though we had been married a year and five weeks, we had cohabited only two months and ten weekend passes.

Now, we had fifteen days to be with each other nearly full time.

Chapter 8:
Peggy and David

Notes from Peg's letters home:

"...February 24, 1944, aboard the Portland Rose...in route to Baton Rouge, Louisiana.

...We left Arlington, 10:16 pm last night...I could tell Dave hurt to leave... He wandered around all evening, just looking at everything.
...We have the nicest compartment.... Boy, this is the way to travel.
...Ordinarily, they serve only 2 meals in the diner, but military personnel get three.
...Train time table...hits right on the minute...Omaha tomorrow night and Chicago 8:30 Saturday morning...2 days and 3 nights from Portland...Visited Museum of Natural History... left Chicago 3:15 pm Saturday on Panama Limited, an all Pullman, diesel-electric streamliner...fresh shrimp cocktails and strawberry shortcake...arrived Hammond, Louisiana 8 am Sunday morning."

Dad drives us to Arlington to catch the train. S*ixteen years earlier, 1928, Dad and five year old me crossed the Columbia River by ferry to Roosevelt and caught the Great Northern, to Chicago. I learned all the way. Finger bowls, Rocky Mountains, Great Plains, six or seven pieces of silver and white tablecloths.* This time with Peggy, I am still learning.

We sleep through Memphis, but are up and going with great anticipation nearing Hammond, Louisiana. I feel immersed in the encounter with outside air; warm and humid. Only in the heated enclosed swimming pool at OSU have I experienced anything like this.

A bus takes us the fifty or so miles to Baton Rouge. The hotel's big ceiling fans are a new and cooling sight for us.

I check in at the base. Kind folks alerted by Peg's Great Uncle Lewis and Aunt Lola have called around locating places we might rent. When I got back from the base, they had taken Peggy to look. She selects a three-room apartment with front porch and backyard. We share the bath with the owner. We like Mrs. Flory; hardly ever see her. Her son is in the Air Force, a gunner on a B-24 bomber. We are there when her son becomes missing in action.

Notes from Peg's letters home:

"...(impressions) bright blooming trees, red, purple and yellow azaleas and camellias...three story mansions with pillars and little shacks...humid warmth...tiny chameleons on the window screen...fried catfish..."

Harding Field didn't have room for all recent Advance graduates. Some would be sent elsewhere. We paid rent for only two weeks.

Class 44-A is held up on weather. We go from February 8th until April 1st without flying. The first flight of 44-B class is March 31st. Over forty P-47s are in the air when the whole area socks in solid. I am shooting skeet near the approach end of the active runway. I don't remember losing any pilots but five or six planes skid, crumple, or scoop into the ground this day.

Planes follow radio homing headings to base. The homing direction is, in this case, the end of the runway. P-47s zip in from all directions. To only mess up a half dozen airplanes speaks well for the training given in advanced school as well as extra cockpit time on those days of weather waiting.

One Thunderbolt comes so close over the skeet range, about a hundred feet in the air. The pilot spots the end of the runway from an impossible angle for landing. As if it were on a carnival gimmick, grabbed by a hook, he pulls hard left and stalls right. The plane's wing partially crumples as it plops on one wing tip, one wheel and its tail wheel in a crunch to a stop. The old bird lands like a pelican's "kersplat." The cockpit occupant slowly exits on rumpled aluminum to ground.

Our Flight Commander, First Lieutenant Schranz, is a big flexible-faced nice guy. He flew combat in North Africa. He says what he wants, then tells us how. We fly a lot; have fortunate weather.

The P-47 is the high altitude plane of the U.S. Air Force, about to take on aggressive ground support as well. We do a little escort training and some altitude dog fights and shoot at towed targets. Like Luke, we are well instructed. I score well on the ground and in the air. We fired at ground targets like old cars and junk. We also fire at bull's-eye targets. They look to be about six or seven feet square. We each fire on an assigned target. I really concentrate, try to do well.

First pass, seems I do fairly good right on range, a little low. I'll raise the sight picture a trifle. The radio lauds the next pilot, call him by #98. Second pass looks pretty good. A few seconds after me, 98 fires, again the controller cheers, "Nice going 98."

I think I am doing fairly good, just want to have my number called at least once. My hits are in there pretty good, I know. I have a couple more passes, but never get a call. I am tired hearing the controller praise this guy. Turned out I had a lot of dumb luck that day, setting a new ground gunnery record. My number is 98. If I hadn't forgotten my number, the pressure would probably have caused me to blow a pass. Great lesson.

Notes from Peggy's letters home:

"...When we were first here, Mrs. Flory had just come from seeing her son off overseas. He is a gunner on a B-24. She and his wife just received word he is missing in action...her only son."

"...I think I'll be home the first part of June...I don't think David will get any time to come home before his overseas venture..."

Every flight, whether dog-fighting around billowing white clouds at 30,000 feet or streaking across swamps on a low altitude cross-country run, is valuable. I, with a wing man, fly north under 200 feet to Shreveport. Our briefing is to keep alert and clear. Our heading is by compass. Suddenly, my wing man drops back. His big four blade prop is flinging leaves and branches wildly during his lucky disengagement from the swamp's gator-like jaws. He had not switched fuel tanks. Its crew chief said its turbo tried swallowing a buzzard, nest and all. The old Jug responded to his error with swift action, gulping gas and scattering brush--helps build our confidence. We fly down the river; looking up to tow-boat pilot houses on the way back to base.

Notes from Peg's letters home:

"...Elections are funny here. They had their primary a month ago, and have been calling the Democratic nominee, the gov-elect ever since. Yesterday was official voting day, but no one voted because the

Democratic primary is all there is. The Republicans don't even have a primary. Incidentally the new gov is Jimmie Davis, a former cowboy movie star and composer of "You Are My Sunshine."

"...learned that a "Mississippi Float" is a big glass of water filled with ice cubes...discovered watermelon pickles."

"...David bought me a dress. I modeled for him...two-piece chartreuse with little buttons down the front...out to dinner and had frog's legs."

Every flight seems exciting. Practice missions with bombers, ground attacks with bombs, and machine guns. Rat racing around cumulous clouds, WOW!

One training flight stands out. We are making gunnery passes at a tow target about 12,000 feet over Lake Ponchatrain. I add a little throttle climbing back to position after a pass and get no response. Nothing happens. No power. The prop is turning, engine is running at about high idle. Switches, ignition and fuel, throttle and prop, all seem correct. It takes me a couple of seconds to conceive it's my Jug, not me, that has a power problem. I am on the gunnery range with gravity alone my source of power. I can see mainly water below or treed swamp surroundings. My estimated range includes one small field between swamp and lake behind a dike.

The P-47 Thunderbolt's glide was about the same as a brick's. For a farm boy, the thought of jumping and consequently destroying that beautiful piece of expensive machinery is unthinkable.

A couple of descending turns puts me over a swamp, crossing a railroad fill, heading directly towards a big barn, wheels up and belly inches above a small farm field. Forward on the stick and the four-bladed prop slices the wet soil. The plane porpoises lightly, then holds together, belly flops once and roots to a stop. I am out and running while it is still settling.

We end in a scrub thicket. I am OK and the Ol' Flying Jug doesn't look too bad for wear either.

Events following, perhaps, gives me the perspectives of both a politician and a bureaucrat.

A familiar sounding little tractor comes from the barn. I checked and recheck the cockpit. My thoughts again are did I foul up; could I have done better? The tractor pulls up. The landowner looks us over pretty good. He is happy to see both me and the plane are U. S. and I am not injured. I finish checking the plane and open the compartment holding the government damage release form. We are well briefed. I am doing the prescribed procedure. The owner eagerly signs the release form which essentially says the event has caused no damage to his property.

I have belly-drug across his harvested field. Its soil is so moist the blades of the prop knifed deep in his field like a giant scimitar. The retrieval equipment will churn axle deep through the soil creating big ruts and whatever other damage it might do to his fences, gates and driveways.

I ride, standing on the drawbar of his Farmall model C tractor, across the field to a house and phone. We go up the wooden steps across the porch and through the door into a large kitchen. The hand-cranked telephone is hanging on the opposite wall, just like ours at home. At the invitation of the owner, I start to the phone, only to be out strode by my host who is exclaiming, "Got to get my red flash signal in!" He lifts the receiver and turns the hand crank with emergency vigor. His measured delivery is staccato-southern, "ONE—SINGLE ENGINE—DOWN—IN CANE FIELD—HAHNVILLE—LOUISIANA."

The big kitchen with friendly folks could be in almost any farm or ranch home in wartime 1940s Oregon. I call the base, thank the folks. Mr. Keller tractored me back out to the pan-caked plane to await further developments.

Several people gather around the plane at a respectful distance. Lots of youngsters, both black and white. I had asked a black Army Private who was home on leave to stand by with me to guard the plane in my absence.

It isn't long until the owner returns with a message and number for me to call. This time when I return from phoning, the Private is having a terrible time. My host's friends and relatives have arrived and are climbing all over my plane. I size up the situation and grasp for an effective solution. I say, "Mr. K—Sir, I was flying and firing in a gunnery pattern when my engine failed. Those guns are loaded. I sure wouldn't want anyone hurt."

With a broad sweep of the hand he yells, "Get the blank blank away from that airplane!"

His block-buster voice performs a vocal concussion-assisted area clear-out. I control my urge to give both a sigh of relief and a wink to the Private. The airplane is safe and so are the people.

I am flying the next day. The aircraft is airborne in three weeks. I sweat out the accident report. It is most important not to have pilot error on one's record. The plane had been in my care.

The hindsight points in this incident are: In the emergency, my instincts were to save both me and the ship. (Came from my upbringing and farm experience.) When safely on the ground, procedure and my career were important to me. (We were instructed to get signed releases – I think this was only the case if the form was not burned up.)

I know that more land and property damage is likely to happen before the plane is retrieved. I am a performing bureaucrat.

Third, the owner remembered his civil defense duties, but he also remembers his family's possible prestige by inviting them over to the accident whereby they swarm over "my" aircraft altogether disregarding the guard, an overruled bureaucrat.

Fourth, I desperately use youthful ingenuity to restore the sanctity of the aircraft and to protect my posterior.

I am appreciative of the kindness and concern the owner and his family show me. They even brought me supper complete with napkin and tablecloth.

Overriding all, is my concern to have not fouled up. Being a single engine pilot has been my goal since childhood. I have earned the wings. Can I earn the assignment?

The plantation owner set his own policy. Perhaps we both use politics. The accident report came out nine days later.

1. The accident. On 9 May 1944 at 1620 CWT, the pilot of a P-47 belly-landed near Hahnville, Louisiana, resulting in major damage to the air craft and no injury to the pilot.

2. The lesson. None. Material failure of the ignition system was the cause of the accident.

3. Action. None.

No sweat. Nothing to it. Just follow the procedure and don't foul up. *Whew!*

Notes from Peg's letters home:

"...Had a bit of excitement yesterday. David made his first crash landing. About 6 pm last night, a Captain called me from the field and said Dave would be late as he had landed in a field and they had to send a car after him. He said he'd be home by 7:30...I judged Dave was all right so I got supper and waited--and waited--and waited...Dave called at 10 minutes to 1 am...I was so glad to hear him. I was cross on the phone...but cried on his shoulder when he got home."

"They got Dave's ship home and it's going to be alright, that is, he didn't bash it up. That is good as he is about the only one who has made a crash landing without really crashing the ship. It's a bit difficult to bring a ship in when it's really hot – 130 miles an hour and no wheels."

"...When I was waiting for David, they played "Comin' In On A Wing And A Prayer" on the radio."

"...saw a redbird...caught fireflies...gardenias are in bloom on huge bushes and smell wonderful at night..."

"...David flies all day now. Yesterday his ship went into a spin at 8,000 ft....he managed to get it out at 5,000. If he hadn't got it out...he would have had to bail out."

"...I really don't worry about him tho, because he keeps a level head and doesn't get excited when things go wrong."

"...He should be OK after this as this is the third thing that's happened to him this week...Saturday his prop went out, and he had to make an emergency landing at the field... "prop going out" means something slips and it goes very fast, slicing the air rather than taking a bite on it. An emergency landing is having the runways cleared...fire trucks, ambulance and flight surgeon on the field...made a 3-point landing OK."

"...David put a dollar on the invasion date...I hope General Eisenhower sees it his way and we win the jackpot."

"...Bought a footlocker to pack our belongings and ship home making plans...train or plane...sold our little Ford 60...Did OK...bought for $575 and sold for $500...three months transportation for $25 per month."

Peg and I took a taxi out to the airport terminal from the Airbase. Our trunk had been shipped home so she had little luggage. We didn't say much. This would be her first plane ride.

Passengers were filing through the gate to the waiting DC 3 transport. Peg looked great. Sunny and sweet. We had very much to remember and solid plans for this and the next phase of our life together. Peg joined the boarding line. I walked out another gate, across the tarmac towards the squadron flight line beyond the runway.

Peg's plane is maybe five hundred feet above. I swallow and smile and feel good. She will be fine. So will we.

Chapter 9:
Peggy

December 4, 1944

From the living room window, I see the car coming up the road towards our house. It turns into the driveway and stops. Two ladies get out and walk towards the front door and I know.

I step into the hallway out of sight and mother answers the doorbell. I hear my name and walk back into the living room. I see only an outstretched hand and a yellow envelope; hear the words, "Casualty message."

I walk back through the hallway to the kitchen and standing by the sink, I open the telegram:

"Mrs Peggy S Childs Dundee, ORE

The Secretary of War desires me to express his deep regrets that your husband First Lieutenant A David Childs has been reported missing in action since seventeenth November over Germany. If further details or other information are received you will be promptly notified.

Witzell acting the Adjutant General

Mother comes in and I say, "It just says missing--not killed."

My father is busy at the barn. I hurry down the path through the bare-limbed apple and peach trees to tell him. He has not heard the car. My Dad is not a demonstrative man, but when I tell him the message, he holds out his arms and holds me close.

How to tell David's father? Lawrence is living alone on Shutler Flat with our little dog, Flicka, for company. We keep in touch by letter and telephone. He tells me of community interests, the Grange, high school games, Sunday school, and, of course, we share whatever news we have of David. When apart, whether school or now in the military, writing letters to each other has always been very important to both David and Lawrence.

My younger sister, Mary Ann, and I had ridden the Greyhound to Arlington in June after I came home from Baton Rouge. Lawrence met us and we had several pleasant days at the ranch, visiting, playing with Flicka and enjoying Lawrence's good breakfasts. No one except my grandmother could do such good fried eggs.

Less than an hour after the telegram delivery, Daddy and I are in the car on our way to Eastern Oregon. The drive along the Columbia through the Gorge has a quieting somewhat mesmerizing effect. My father may be thinking of trips he has made along this highway going hunting in the Blue Mountains. I am remembering the first time I saw the Gorge, almost two years ago when newly married, David brought me home to meet his father. That was the time of nervous expectation, this time, I am dreading delivering the message I have. Daddy and I both notice and remark about the great jutting stone bluff just before we cross the Deschutes River.

It is dark when we drive through Arlington and on out to Shutler Flat, past "the tree." We stop at a neighbor's ranch house. I tell them why we are here and ask if they have talked to Lawrence in the last day or so. They have and say he is well. As I leave, the lady of the house says, "How nice to see you."

Seems a bit odd considering the purpose of our visit, but I know it is well meant. We drive to the crossroads and then the last mile to the house. I can see the light in the windows. My father had visited Lawrence in October on the way to his annual elk hunt in the Blue Mountains. He had brought jars of peaches mother and I had canned during the summer.

We park under the trees at the front gate, walk around to the side door, knock on the inside door and enter the living room. Lawrence is sitting in his big wooden rocking chair, the one with the high back and extended legs. He is wearing his glasses, a newspaper in his lap. He does not seem surprised to see us, as though he knows what we are going to say when we walk in.

My first words are, "David is 'missing in action'," and I hand him the telegram, adding as quickly as I can, "not killed, just missing."

The three of us talk, trying to reassure ourselves with all the probabilities...bailout, walk out, prisoner...but we know, just know... he is alive.

We talk about other things, visiting--the weather, how Lawrence's chickens are doing, casual easy things, but the specter of the telegram message is always there.

It's time for bed and Lawrence sees that we are comfortable. I kiss him good night and go up the stairs to David's room where we had slept during the almost two months after we were married before he was called to active duty. Everything is just as we left it after our visit home last February, after David's graduation at Luke Field. The heavy quilt of pieced woolen squares tied with yarn on the bed; David's school and athletic trophies are on the walls; the cretonne curtain I made to cover the closet shelf David made.

It is cold this December night. I find David's old plaid shirt, wrap it around me and try to write a letter to him--telling him how it is here and how my thoughts and feelings are with him. I look at the telegram and realize it says *First* Lieutenant. He has his promotion and I can't tell him how proud I am.

It is so quiet tonight on Shutler Flat. I miss the wind's singing through the power lines west of the house. Then I hear heartbreaking sobbing from the room below.

It is morning and Daddy and I leave for home. Lawrence tells us goodbye with cheerfulness and hope in his voice. We assure each other David is strong and capable; he will make his way back.

Lawrence makes his own journey east to Pendleton to the State Hospital where David's mother is now living.

She sometimes, though rarely, is aware of things we do not know.

David writes--

Dad described to me, his seeing Mom in this manner, "I came to the open door, Cavy was standing by the window across the room. She turned looking directly to me and said, 'L W, our boy is in bad trouble, but he'll be all right'."

Chapter 10:
Stalag IV-B

A night and day; a second night on a hospital train. I have no idea where. I am told, "Leipzig, then you travel another hour further to where you will be going." Someone else speculates Dresden; said it is noted for medical training. My farthest mission into Germany was forty miles northeast of Frankfurt. Another mission with rockets was northwest to Aachen. These missions, both quite successful, would not be recalled until after this one plays out.

Orientation for us in ground attack was confined most of the time to our sector. We knew the names and worked with several ground controllers. We operated mostly in the Third Army sector, however I flew on a squadron mission to destroy Tiger tanks with rockets east of Aachen up north because of our squadron's rocket capability on dug in tanks. We had a bang up mission northeast of Frankfurt, breaking rail tracks and an encounter with ME 109s, however, most of our missions were west of the Rhine; Trier, Metz, Saarbrucken, Kaiserslautern areas, trying to deprive Patton's opposition their supplies. We continually broke track in multiple places. My knowledge of Germany was meager.

I wonder about my whereabouts and mostly about improving my recovery. My condition is worse than I first thought. So far, I have been extremely fortunate. I am determined to take whatever comes in the very best way I can. I am coachable and need a coach.

I do know Peggy, my father, our families, and the folks back home; Condon, Arlington, Lone Rock, Olex, Mayville, Igo, Clem, Mikkalo, Rock Creek, Blalock; the whole county is pulling for its "boys."

I do not tell anyone anything about my missions, outfit, or what I flew. I am sure my German captors have answers on file, but I do not have many for myself. I figure from fragments of conversation heard after my ride in the open railroad car that I am being carried by POW laborers to a POW camp.

I arrive about afternoon snack time according to my internal clock. I can hear questions and replies in languages I cannot understand; directions in German, French and British. It seems 'rather' efficient, and apparently I am not to be a terrific blip in the scheme of things. Shortly, I am delivered to a compound in a form more like a fire log or a butcher hog than a bold American airman. I am tilted through an entrance, then carried down a hall and into a room. My delivery people, now down to two, place me in bed, bidding a figurative *au revoir*.

I am settling into a straw mattress, wondering a little and hoping a lot, when this kind British voice offers a warm welcome. "Hello, Yank. I'm Jack."

I'm sure he sizes up the plight of this swaddled log or hog and starts right off with facts. He is testing, learning about me, and shows merit and skill.

"You are in Stalag IV-B. It is a British NCO (non-commissioned officer) camp with other compounds including RAF, Russian men, Dutch, French, and one of Russian women. We are on the Elbe River near the town of Muhlberg. It's tea time. Would you like a cup of tea?"

"Sure," I say.

He holds a cup of tea for me, steadies it with my hand. I am thinking my bandages are grimy, and the edges of the mouth opening stained and not too sanitary. The ones on my hands are also the originals.

Jack continues, "You are in the medical center for the camp. It's a compound fenced in from all the others. The staff members are French prisoners of war. The principal doctor and surgeon is twenty-six years old; fine man, Dr. Robert Monnier."

I don't ask questions. Jack knows I have been down twelve days, and this is my first experience with POWs in a POW camp. He gives me a Canadian biscuit covered in strawberry jam. I think to myself, in my best British accent, "What ho, this is really alright!" The way Jack says "Canadian biscuit." I recognize this is a special treat. The way he pronounces "strawberry jam" makes it sound to be sweet, fruity, thick and delicious. It was later I learned some Brits like orange marmalade as much as I like strawberry jam. Each think "I got the best of the trade."

Stalag IV-B

Jack tells me about food parcels; usually one parcel between four people each week. (They were meant to be one for each prisoner per week.) Jack says the German ration of bread and margarine is already divided for the day. They will share a little for me for the evening meal.

Jack covers a lot of ground in a short time. I like him. He can coach. Just what I need. (He and another Jack, one who follows him in the ward, plus Ron Bleecker, and eventually another American P-47 pilot, Jene Knierim from St. Louis, are all enthusiastic men who improve the chances for me to not only survive but to rapidly improve.) After the tea is brewed on a little briquette stove, it is poured through a tea leaf strainer into each cup. *I hear the process and discussion.*

Jack says, "You have belongings that came with you."

I ask, "What did you say?"

He says, "There is a paper bag with your belongings that came attached to your stretcher, also your combat boots. Would you like to hear what is in the bag?" I say, "I sure would." I am curious and surprised.

He starts setting things out naming each. I feel them touch the bed. "Shoulder patch, U.S. Air Corps collar insignia, fountain pen Parker 51, cigarette lighter Zippo, New Testament, says Peggy Shelburne." (Given to me by my wife from her Grandmother Shelburne. I always carried it in my shirt pocket.)

I feel gratitude, amazement, and pleased. Jack continues, "A thing like a lighter with a six inch wick on it." It is a kind of cigarette lighter. I carried both lighters. I'd planned to survive if the need arrived. "Your identification disk."

I interrupt, "What?"

"Your broken chain and I believe you Yanks say dog tags."

This news really stimulates thoughts; the chain was broken a couple days before the mission in a volleyball game. I tied the ends together with string. I couldn't find the dog tags when first downed. My burns are more troubling, especially around my neck. I guessed the string either broke when the chute opened or, more likely, burned through and the chute opening jolt had sent them down into my underwear. I did a cursory search after I missed them. I figured they were lost in the snow.

I chuckle. The German sergeant did his job. The interrogator tried the ruse. His training and my training ended in a push. I knew nothing.

He didn't get me going. He was like a geologist with a rock pick hoping to find something, but just exploring. I expect they checked this brown eyed, dark haired man named David for most everything even before the interrogator arrived at the hospital.

Jack seems surprised finding two wrist watches and a book of matches stamped with Morse Code V...-for victory. He says he will get me new clothes when I need them. They give me hospital pajamas and my personal urinal. Burns seem to promote drinking lots of fluids. I am unable to walk but my legs feel better. Jack tells me I will need a spoon and a table knife. He asks if I smoke. I tell him, "No." I had a pipe but seldom smoked it.

Jack says, "Good. You will get a German ration of thirteen cigarettes a week. We can use three or four to buy your utensils from the Russian with the great greatcoat. He will be around."

I give everyone in the room a cigarette. Jack says it isn't necessary. In order to make food go farther, I am to be partner, or mate, with Jocko, a Scotch RAF Sergeant from Aberdeen. He has a leg in a cast. By sharing, our food will go farther with more variety. Half a parcel for two people is better than one-fourth parcel for one. Jocko was a Halifax gunner or maybe a Sterling. Both aircraft are mentioned. Like I say *Ore-gun*, Jocko has a special love detected in the way he says *Aberr-deen*.

I sleep pretty good. The urinal becomes an important companion day and night. People help each other. I listen a lot. There is always someone talking until late.

Jack's name is Lomath. He is in his late twenties, a Londoner. He'd been a chef in civilian life as well as a chef in North Africa. He jokes about his hospital stay, says he had a near nervous breakdown during the camp's fall theatre presentation. He had a tough job in the production, a London musical; says he'd come into the infirmary briefly and Doctor Monnier kept him on as morale booster. He won't be able to stay much longer. A German doctor comes around weekly to check for compound prisoners overstaying.

The second day, I am taken into the dressing room on a gurney by two small Frenchmen. Dr. Monnier looks me over. He consults with a British POW doctor. The small Frenchmen spend a lot of time debriding my two weeks old burns. The bandages are pulled off every other day, a tedious "ouch" restraining event, takes about two hours the first time. I

Stalag IV-B

am pleased because I can see a little. My eyes are goopy. I can't see the walls or floor or much more than blurred forms.

Seated in a barber like chair or maybe a dental chair, I see enough to observe the large deep burn on my right thigh. It hasn't been particularly painful. It explains the rows of black burn holes in the fanned out parachute's canopy draped over wild rose bushes seen on landing. A 20 mm incendiary projectile likely penetrated the seat-pack parachute and was nearly spent when it wedged in or against my thigh and flight suit, burning the parachute and my leg during the seconds it took to get out and pull the rip cord which removed the phosphorous culprit.

The sequence of happenings was very fast. I jumped clear in approximately six seconds after seeing smoke. This is about five seconds after feeling two hits. I dwell on the mechanics of leaving the plane, both as a mental exercise and to satisfy curiosity.

Not having vision, no method of recording notes, and no tables or consultants, I have little to go on other than knowledge gained in physics classes at OSU and cadet ground school.

60 mph=88 ft/sec; my speed 400 to 450 mph;

100 mph= almost 150 ft. per sec x 4 = 600 ft plus for each sec.

That's football field, goal line to goal line, twice in one second.

Altitude above ground when I jump: At or under 200 ft.

I visualize the sequence: Chute opens, the Jug fire-balling into hillside, evergreen treetops interrupt view of Jug on ridge ahead, a half length of our 5000 ft runway airstrip at A-79. Looking down, I hit hard in less time than it takes for a goal line quarterback sneak across.

The mathematical limits of speed of bailout, falling distance and distance traveled don't add up in my wrapped and swaddled head. These are facts I determine:

1. Both thumbs were black as if hit with a hammer, before I moved away from the parachute.

2. The GI knit wool liners to my burned gauntlets were battered; the left little finger stall had the end torn away.

3. Ligaments in both knees were stretched so I had to balance to keep my thigh bones over the leg bones. Jaw was stiff, hinges strained.

4. The parachute draped over wild rose bushes, showed at least two and possibly more rows of blackened holes across the canopy. Holes big as tea cups, smaller than saucers.

5. Escape kit and its pocket on right leg of flight suit, are ripped off and gone.

6. My GI combat boots have both laces and buckle straps. Laces were knotted tight. I couldn't get them loose to snug up. They felt very sloppy on my feet.

7. My wool trouser pockets held some items, but my stockman's knife had escaped. There was a hole as big as my finger in the pocket.

8. My dog tags were missing from around my neck.

9. My forehead felt like an over-cooked Hubbard squash in the ranch kitchen's oven.

10. My left eye was closing. My right eye seemed to respond to snow bathing.

11. My ears and nose were seemingly OK.

12. I shoved dark goggles up when black smoke filled cockpit, same time I vented the canopy, instant ignition. I figure my flight helmet protected my ears until the canopy jettisoned.

a. My eyes were unprotected at ignition.

b. My shirt cuffs, shirt collar, gauntlet-gloves protected somewhat, but the stretched openings exposed hide when I jerked the canopy-jettison pull-cord and thrust feet down hard in attempt for a clear-the-tail body launch.

Speculative:

a) Burns on calves of my legs possibly occurred prior to seeing smoke.

b) Switching on full oxygen may have been a plus or minus. The mask flowing positive rather than demand oxygen prevented smelling first smoke.

c) Dark sun lenses in goggles, changed lens because of white snow and forecast of sun breaks. We'd been weathered down for several days; our specialties were needed.

13. It is easy to second guess what happened wrong; why the two locomotives were not both taken out the first pass.

Only recently, I learned P-47 bubble canopies were suspected of striking the back of the pilot's head, stunning him fatally, when ejecting the canopy at very high speeds. (Letter from Adam Quandt, former

Edwards Air Base Tower Controller.) Maybe my venting the cockpit prevented the canopy from doing the fatal caper described below.

Adam Quandt is a highly regarded 377Th replacement and former Thunderbolt instructor, who joined my mates shortly after I was downed. Combat friends admired his pilot skills and first rate personality. I met him at his first reunion with the squadron at Salt Lake. I soon learned my buddies were right on. He seems a great person. I look forward to more time talking with him.

Letter to Dick Jones, close relative of Archie Billings who was a fine friend in our squadron (Archie is mentioned later in my learning of his loss).

To Dick Jones   August 19, 2001
From Adam Quandt
I have finally found my mission records, and I can tell that I was on Archie's wing on the mission he was shot down. I will send a copy of the mission report later, but feel I have to add something to it now. The flak was very heavy and accurate. He and I dove down to strafe a flak battery and I saw Archie's plane burst into flames, and yelled at him to bail out because he was on fire. He never answered, but the canopy came off and I expected him to bail out. He never got out of the plane and dove into the ground almost on one of the flak positions. Sometime, maybe a month or two later we learned that when the canopy was ejected, instead of going off smoothly, the leading edge tilted violently downward into the cockpit and this conked the pilot on the head and either killed him or knocked him unconscious. A hell of a way to go. If I can be of any further help, please let me know.
   Adam Quandt

Chapter 11:
The Ward

The first trip to the dressing room gives me an inkling of my condition. I can see a little. My face is burned badly around the eyes and the entire forehead. My right leg is deeply burned into the flesh of the thigh, third degree as learned in first aid. Two French POWs, former dental assistants, tweeze and soak and peel and pull at two large areas of no man's land-like destruction. The backs of the calves of my legs are only first and second degree burns. The backs of my fingers, around my neck, and where the oxygen mask stopped under my chin, as well as around both wrists, are also third degree. I wore a watch on each wrist. The skin around my eyes, chin, and thigh take a lot of care. The fingers soon improve.

Room living: The routine is ad lib, somewhat military because it is a soldier hospital, about three wards.

British NCO Camp: Most were captured by Rommel's troops in the wadis and sands of North Africa.

Ambere': A French prisoner male nurse. He is witty and patient, probably twenty-four or five. He has a sense of humor and a serious side as well. He is practical, but isn't above a little joke now and then.

Dr. Monnier: Tall, aristocratic, twenty-six year old French POW. He has a fine presence, no fanfare; diagnoses frost bite, pneumonia, hemorrhoids, nerves, appendicitis, fractures, infection, and intestinal ailments. The ward doubles as his examination room. Ward mates know first. He has a wife and their two or three year old daughter.

Dutch barber: Uses shears, straight edge razor and skill. He preps all the surgery patients and does occasional shaves and haircuts.

Jack Lomath: An Englishman, around twenty-eight, kind and cheery. He is a trained 'French' chef, practiced in the Savoy in London. He once fooled the Prince of Wales who took him for a Frenchman.

The Ward

Jack Thatcher: Tank Commander, about twenty-five, a dapper London man about town, responsible and likeable. His family are undertakers in London. The business, known as Toukey's (spelling guess by memory's ear), has been in the family for two hundred years. They also ran their casket factory and maintained a stable of a hundred carriage horses, renting horses, coaches, and carriages to mortuaries all over London.

Sergeant Major Sam Ashmore: Senior NCO in IV-B. I sense respect for him by all.

Major White: A British doctor, is one of four diverse nationalities who play bridge on occasion. The four speak Russian, French, English, and Dutch.

Corporal Ferguson: I believe he is called the British Man of Confidence, similar to his freely moving German counterpart, *the ferret*; his name is Sperling, something like Sparrow, anyway, the "dapper" Jack carefully refers to him as "Dickey bird."

There are four chaplains. The British call them Padres. I memorized their names and sometimes I am cued with a couple notes of "By yon Bonnie Banks" hummed by Jack Thatcher when Padre Banks visits the ward. (I wrote their names down on an envelope in 1945, or I would have forgotten the three other Padres; McDowell, Willis, and Day.) They visit the ward weekly.

The care is great. Both Jacks explain what is going on, so when I finally can see the surroundings, it isn't much of a surprise. To the right of the hall door a small stove burns briquettes for tea time only. My bed is in front of the door to the hallway with two more beds to my right. I believe three on my left. Jocko and I are the two newest patients recovering from battle damage. Others are victims of wear and tear, or maybe pneumonia or appendicitis.

We are brought a kettle of ersatz coffee shared in the mornings. Breakfast is a piece of black German wartime bread with a swipe of margarine and a little ersatz jam, supposed to be made partly from coal dust. They say it looks like thick ketchup.

Tea, mid afternoon, is best of the day. Someone starts a briquette fire in the little stove and heats water. I don't see the ritual but hear the sounds of stove top and fire making. The water boils until tea is just right. The tea is bulk leaf and is poured through our room's strainer.

There is small talk with each step in the process. At first, a cup is held for me. I guide it, gaining more confidence. Within a week, I hold the cup and cracker and sip and chomp away.

By the time I am able to see the preparation, the room is pretty well Americanized. Most of the Battle of the Bulge POWs came through Stalag IV-B. (In the last weeks leading up to my leaving, one eye was left unbandaged). I arrived November 28th when the room was pretty stable. Jocko and I become accustomed to each other. He divides our ration, collects our cut of the food parcel, and in general, prepares our food. We trust each other. I am entertained by the Brits, educated too, at least learning a lot I never even imagined. Jocko is prone to mention the Halifax aircraft and flak, but the old hands from North Africa joke about some tough encounters there. Best and worst was when the Italians tossed in the towel. They were in transit in Italy. The "Eye Ties" turned their prisoners loose, opened the gates or doors. I can't remember whether they were in a compound or boxcars. Jack Thatcher described it. "Whoops. It felt good. We were free, even cheered."

Then like a sand storm on the desert, German trucks, machine guns, a halftrack, and a Brigade force of full battle-equipped Wehrmacht swept them up and closed the lid. Their elation became dejection. They turned to outwitting their captors for recreation. Both Jack L (Lomath, the chef) and Jack T (Thatcher, the dapper one) made it clear. "One does not ridicule Jerry. One does not laugh at Jerry."

The wall running behind the heads of our beds has several windows. I have been here a little more than a week, when Jack T points out the rumble of armored traffic moving by the camp over several nights. Jack said, "Jerry has something going."

We live somewhat like sleepover kids, talk goes on at all hours. Lots of tea and a chilled room make for realignment in bed and the breakout of conversations at all times of night. I usually ask what time it is. Ron Bleecker thinks it uncanny that I can tell within a half hour what time it is in twenty-four hours, without seeing. Actually, my ward mates tell me the time with their day and night actions or doings.

Today, the Dutch barber is in the ward for haircuts and shaves. Jack Thatcher explains the scene with some verbal pantomime. A Brit is being shaved which includes trimming a neat mustache. Apparently the barber is lathering his face when the lad questions the barber, wanting to know if the Dutchman uses the same equipment for barbering that he uses for prepping patients for operations. Two or three of the senior patients begin to quiz him about preps of the past while the barber continues the delicate trimming around the recipient's mouth and under his nose. Of course, I couldn't see any of this happening, but my imagination is working fine.

I have firsthand knowledge of barbershop shaves. It became a harvest time Saturday afternoon ritual for me and the driver of our sacked wheat hauling operation to stop, last load in town, and get a real shave and trim. I was sixteen.

I could picture what was happening. Jack was my interpreter.
Question: "Is that how you prepped John's appendix op?"
"Yep."
Question: "And Jocko's leg, broken just below the pelvis?"
"Yep."
Question: "And Jack T's hemorrhoids?"
"The same." With a couple guffaws.
Question: "The Russian woman's...?"

The barber chuckled. I grinned under wraps. I felt for the current patron. The point was clearly made. The Dutch barber's shaving brush had been around.

Some nights, engine and tank tracks noise was unmistakable. Tanks were being moved. I could hear them through my bandages. Jack mentioned it often. "Jerry is up to something."

Christmas is coming. The room, as a whole, starts collecting bits of parcels for the Christmas cake. The cooking staff will give our room chef space on the stove in the kitchen. Our meager ration of coal is only good for tea time. Winter weather reminds me of upstairs back home on the ranch. Our ward is just a little above freezing during the day and colder at night. Most spend much time under covers, night and day.

This morning, I can tell some sort of skullduggery is afoot. I depend on my other senses to make up for not seeing. Probably the sense of deduction is the keenest. Monsieur Ambere' stops and stands at the foot of my bed. (Something he never does.) I can feel the attention of everyone in the room; there is a noticeable silence. Someone is grinning out loud. Things are different. I have never seen Ambere'. He never comes into the dressing room while I am being worked on. Seldom does anyone come in other than doctor and his two helpers. I like Ambere'.

He never complains and he seems, to me, clever and resourceful. I kid him about my French language skills, which amounted to sounds like "Chevrolet Coupe' (cooo-pai)." Ambere' would say, "Whas e say Chevrolet cou<u>paeee</u>?"

I would say, "Bon France."

He would come back, "No, is *par bon*."

Ambere' begins, "*Monsieur DIE VEED*." He was reading. I can hear a paper crackle. "TAKE YOUR HAND OFF EET!"

I check my position. Sure enough, I am lying in bed with my knees bent, my rump in the hollow of the straw mattress, with my hands folded over my crotch. I think this is the standard energy conserving, protect the future, cold as a wedge, survival position. I wonder how long I have assumed the blinken', bed ridden, prisoner of war, gotcha' pose.

Jack T asks if I'd like a bath. Says he can do it. He gets water and begins scrubbing on uncharred body sections. Feels good. He is in his jovial mood, alternately humming and talking. He bursts out, "Divey, me boy, you're me first live one I ever did."

Before Christmas, the German ferret makes a daily appearance in our ward. He says the German Ardennes attack has taken the Americans by surprise. Next day, same thing. "Those Americans are running and retreating over a wide front."

I am not very smart. There are only a few Americans in the camp, and I am obviously a propaganda target. He is getting pleasure by needling me. I am thinking there isn't any way the Germans can sustain a drive or hold a line against the many tactical fighter-bomber groups of the Ninth Air Force and the Mosquitoes of the British.

I ask Jocko, "What's the weather like?"

I hear his bed squeak and straw rattle as he strains to look out. His judgment is, "Yank, it's socked in, low clouds."

This goes on for a couple more days; low cloud weather reports from my partner, the Scotsman. The German ferret is getting bolder and more taunting.

I have a lot of esprit de corps. Fighter-bomber ground attack is something I take pride in because we wield a double-edged sword. We trouble the Germans mightily. It is clear to me a squadron of P-47s with a combination load, twenty-four frags or five hundred pounders, ninety-six 50 caliber machine guns with 350 rounds per gun can obliterate much of an armored column, camouflage and all. It puts them on the defense, protects and cheers our ground troops; makes us earn our keep.

I have been reasoning when 'Big Mouth' comes in the next afternoon. He starts his spiel and adds ridicule. He is talking to a room full of British and baiting me. He is on a roll. Says, "You're going to be backed into the sea by those cowardly Americans."

I am not proud of my move. In fact, I feel stupid that I allow him to goad me to an attack. I pull my legs up, throw aside the cover, jam my feet on the floor and start for him and the door. I don't catch him. My legs are locked, won't bend at the knees. I am crabbing forward. I don't go down. He is gone. I work back along the bed in a crouch, roll in, pull my legs in, cover up, and call for Ambere'. No one says a thing.

Ambere' soon comes in. I ignore his mirthful tone when he asks, "What is, Monsieur David?"

I say, "My legs won't work. Can't bend my knees."

Someone explains, probably Jack L, what I am saying. Ambere' understands. Soon a man comes in and tries manipulating my legs. They won't straighten. I get him to bear his weight down on a knee. Neither budge or it doesn't seem they do. I begin to work on them daily but don't try walking except to the dressing room every other day.

Next morning, I am still a little bit razzed. I question Jocko.

"How's the weather?"

He stretches for a gander and replies, "Looks better. Fog's lifted."

In an hour, the sun is coming through the windows warming the room. I am invigorated. I ask Jocko if that is really sunshine. He says, "Yep, it is, Yank."

I am encouraged. I say, "The Ferret won't be in this evening."

He doesn't come again. I think, "Good-by Dickey Bird."

Christmas is at hand. The room is cheered. Christmas presents will be distributed. We will share "Christmas pud" (pudding) and cake and tea. Men from other wards circulate through the hospital. Jack L arranges for me to get a smoking pipe for Christmas. He negotiates for some pipe tobacco, but is unable to close a deal. I don't know what they acquire, but Jack T loads it up and hands it to me. I puff while he fires it with the Zippo he has loaded with petrol from the potato truck on its delivery.

Jack is too valuable to risk in that sort of simple heist, but there are plenty of British volunteers who need the practice. The trick to something either simple or complicated is to confer and rehearse all phases, including both the diversion and the caper. I am not a smoker, but I tell them I occasionally puffed a pipe in college. Peggy thought as a romantic and had suggested a pipe went with a casual jacket I wore. I had not used more than a half can of Sir Walter Raleigh in my two years in the Air Force. I appreciate their gesture a lot and attempt to be cool.

Jack T says, "Look! Smoke is coming from his nose."

His tone is encouraging.

Ron Bleecker comes into the *Lazarett* (hospital) for a few days shortly after I arrive even though we left Kaiserslautern the same day. He is assistant at one of my infrequent BMs. Maybe this event spurs his recovery. Ron is sent out to the main camp. He is kind. He asks one of the Padres to arrange for him to visit me before he is sent out to work camp. My bandage is off one eye when he comes to the hospital. He is a significant youngster.

Later I try to find him. My letter to him doesn't come back.

Fifty years later, 1997, his daughter finds me for Ron's Father's Day gift. Peg and I and Helene and Ron have exchange visits; Michigan and Oregon. Great people.

Soon after Christmas, the ward is full. Two or three beds are added along the hall wall. There is still room for the box by the door which holds pages from the current propaganda book supplied by the Germans. POWs read around the room, one page at a time, first man tearing out a page, reading it, handing it on. Last man drops the page in the box. Pages are retrieved for latrine use. I can't read, but the very meager diet doesn't demand a lot of occasions for paper use either.

On the ranch, Ed and I summered in the homesteader cabin, now a bunkhouse. The outhouse was behind the bunkhouse; had a Sears Roebuck catalog nailed on the wall. During operations, we learned to crumple the slick pages several times with a final rolling between the heels of the hands in order to velvetize a page into effective texture.

Propaganda pages are not as difficult as slick catalog pages. Before long, I join the ward's Reading and Comfort club.

Jack L says his goodbyes and goes back to the compound. His

The Ward 73

cheerful, purposeful and talented manner is missed. We enjoy Jack T's equal talents in his own inimitable way for a few days longer.

Jack T's stories also entertain. He, too, does lectures around camp. One is about the undertaking business in London. Jack often drove the hearse. He and his father prepared King George V's body for burial. Another of Jack's stories concerned Lord Jellicoe's funeral. Jack, nineteen, was chastised by his father for his appearance in a press photo. "I was holding Hitler's wreath in front of me while I peered around it, mouthing the words, 'From *Adolpho*,' with a big smile on me face. Dad said, 'An undertaker or his crew must never smile'."

After the war, Jack's whole family, father, mother, brothers, sister, nieces and nephews and he and Terry, his red-haired, Irish colleen wife, all moved to the West Coast of Canada. The folks sold their business and turned to security, airliner pilots, other things. Jack related his doctor confided, "Jack, when your family came in, they doubled my practice in one swoop."

I asked Jack to bring the laughing photo of him and Adolph's wreath when he visited us on one of many occasions. Jack kept his wit and humor, great showmanship, and warmth. We and our kids loved him.

I have a great Christmas and am beginning to get adapted to the routine and the people. I know by the reaction of my burn debriding team, they think I am improving. The sticking and painful to remove gauze pads covering my forehead and thigh are beginning to shrink a little. I am hobbling to their dressing room with just a little assistance. My knees are flexing some. I am brightened when they complete tweezing and scraping this time. They are expressive and cheerful. In fact, recently they show me pieces up close to my eyeball of material as they pull it off, especially if I flinch. They assure me it is "no good." "Par bon, Monsieur."

This time one hands me a mirror. It is small. I move it a little. My mouth and nose look OK (protected by the oxygen mask). I move the glass, focusing above to my eyes. I am jolted, am not prepared for the image. My sight picture is still vivid after sixty-five years; a red mass, swollen, pulp like, broken by two slits. I couldn't see pupils or eye balls. Nothing like brows or lashes. I had never thought of them. I could only think of a red-faced ape that had been attacked by a tenderizing meat hammer.

My first reaction is, "Peggy, how can I bring this home to you." Tears feel good. My eyes can water up. Thinking of Peggy helps. Her face is there in the tears. I relax. The whole interval lasts less time than my bailout. She will be tough, tender, and there. We can do it. That few seconds is as close as I come to despair, ever.

The mirror is gently taken from my hand.

Chapter 12:
Jene and I

Just before New Years, Jack L comes over saying another American is in the hospital. They will move him into our ward soon. He continues, "He's burned like you only he's much worse and in severe pain."

Jene Knierim is also a P-47 Fighter-Bomber pilot same as me. His legs, face, and hands are burned. He has no broken bones. Again, time soon makes a big difference. Turns out his burns are more painful than mine but not as severe. He isn't very big. The attendant picks him up in his arms and lays him on the bunk's sheet. They seem to be airing his red raw looking legs. I have partial sight. His bunk is by the door opposite from the "book paper box" near the foot of my bed. Jene's recovery is relatively rapid.

Jocko is sent to the RAF compound and Jene is moved to Jocko's spot in front of the window next to me. We take turns fixing food. Once I ask, "What do you want on your bread?"

He responds, "Surprise me, Dave. Surprise me."

Food parcels are stretched further. Conversations gradually are omitting women. Hunger joins Cold as a constant companion.

When my forehead is fairly healed and the swelling is reduced, I may have taken another look at myself in a mirror. I digitalize my eyes and lashes, or maybe deny the visual. It is my expanse of forehead that has changed, with the complete disappearance of distinct vertical squint creases. My hide feels thick with scar tissue. My "perplexed look" is modified. Brow muscles don't work. I have no eyelids. No eyebrows; and no lashes on inner half of both upper and lower lid lines. The corners of both eyes are webbed and harbor gunk. I cannot close my eyes.

The ward changes rapidly. Jack Lomath is discharged from the hospital shortly after Jene arrives. Jack Thatcher takes over for part of Jene's and my care.

Youngsters from Battle of the Bulge and the overrun of the 106th Infantry Division soon overfill the wards. The staff continues care of our burns. Jene is sent to the Main Camp around the first of February. Jack Thatcher takes Jene in with their combine. They are really well organized. They receive parcels from home. It seems certain Jene and I might be moved soon.

With the wounded of the 106th and the removal of one eye's bandage, I can partially see the changing surroundings. I want to help. Even though we are now mostly Americans, we still have tea, though not so fine as under with the British.

I know in my limited way we will likely move. I need strength and stamina. With a little direction, I start walking the perimeter of our small compound, about the size of an American football field. I get better, still can't straighten my legs. They are improving. The circuit doesn't take long, maybe a half hour. I also start going after the briquette ration twice a week. I don't see well enough to play cards, but we play Monopoly. It is a British set. Some of the names I had learned when on combat leave to London; Trafalgar Square, Piccadilly Circus and others. I enjoy Monopoly. Some luck, some skill; kind of like farming. Sometimes boom, but it can bust.

I read a German distributed British novel, "This Above All." I can't see much propaganda value in it. I am told the description of the "brave young German pilots," crossing the Channel had something to do with its value. Anyway, I am first reader and tear off the pages and start them to the box by the door.

I am walking in a bent-kneed gimp but am beginning to go more laps. I have been fitted with a British Uniform; blouse, trousers, and shirt, as well as under garments and socks. I keep my own combat boots. Remembering back when Jack L took my burned shirt, he said, "You will get a uniform when you need it." I had asked him to save the tail of my shirt. We have a saying at home, when a person slips through a tight situation, "He got out with his shirt tail." It means he didn't have much more.

Ambere' comes around with a bottle of cod liver oil and a spoon and distributes it to some of us. I didn't know it is only for those Dr. Monnier thinks needs it most. One day, I kid him. "Where is the cod liver oil? We are hungry." He shushes me, indicating, "You have big mouth." I learn they are metering it out. I get it not because I deserve it but because I need the vitamins. I eat the cheese ration. Some cannot. One cheese is white and soft, it smells awful. Another is hard, octagonal-like, kind of yellow and orange, referred to as "dog turd" cheese. I eat mine. I know we are way short on calories and can use the energy.

War news is better. I still get dressings every other day. The first or second week in February, we can sometimes hear Russian or German artillery, or both.

Dr. Monnier does a couple necessary procedures for me, saying further surgery will be necessary when I get home. First, he takes surgical scissors and slits the corner webs of each eye. Another day, he comes into the dressing or ops room and has me lay flat on my back. My legs still won't come down flat on the table. My knees have taken a pretty firm set. I think my gait is like that of the Hunchback of Notre Dame. Before this session is over, the legs, knees, ankles, and toes will be more flexible.

Doctor holds what seems to be a styptic pencil. I use one in my on-the-go shaving kit to stop bleeding razor nicks. He says it will be painful. It is! He glides the peeled stick flat, pressing back and forth, over the keloids under my chin.

I concentrate on keeping my shoulders firmly pinned to the table. I anchor them but my legs won't hold still, neither will my butt. Dr. Monnier just calmly keeps a steady pace, then moves to my forehead. Man, I squirm, but not my head and shoulders. Finally he stops, saying, "It is enough today." I know he has at best half of the forehead yet to do. I ask, "Do you have more?" He answers, "Yes, Monsieur David. This is enough for today."

I plead, "Doctor, let's finish it. I'm doing OK."

He completes the task, and I, figuratively and actually, get a bottom down wiggle workout. My compatriot POW doctor has done what is necessary. I know it and am grateful.

I feel confidence with him. I listen when he diagnoses and treats. I glimpse some of the blackened feet, hear frozen toes drop into a white porcelain basin. I wonder what will happen to some of these kids. One little Lieutenant with pneumonia is in the far corner. All that can be done here is being done, which isn't much. They have put drain tubes through his back, into his lung cavity. He is so weak. We possibly heard the tanks that got him go by here at Stalag IV-B.

Doctor exams an appendicitis patient seemingly using a gloved digit. Some former observer or recipient of the rectal probe remarks, Doctor Monnier's long finger apparently is a standard examination procedure for various illnesses.

Earlier, Ambere' came into the ward and cheerily walks up to the side of my bed. *"Un, deux, trois, quartre."* (1, 2, 3, 4)

I ask, "Waz zee say numbers?"

This is still when my eyes are covered. With Jack L as interpreter, I found that at last I am going to be of some use. Ambere' needs to test four rectal thermometers. Would I be willing to be the guinea pig, so to speak? I'd had my temperature taken daily, night and morning. Never had this method before arriving in Germany. However, my 'modest' sacrifice will be of benefit to the welfare of Allied troops. I consent. This task is an undercover job. My expertise with manipulating the delicate probe has been proven over several weeks, sight unseen. Ambere' acts as recorder. He doesn't want to get involved in a "whose on first" routine. I insist the proper registration, identification, and thermometer shake-down, turn-over from recorder/nurse to patient/technician be precise with military bearing. Lomath chuckles. Ambere' is gleeful, and Thatcher urges on. I don't remember the variations in readings, but I carefully noted the outcome of each device. I probably could have been presented the Glass Bulb award with Mercury Cluster.

While we are in this area--our sometimes diet of boiled dried peas, stem, pod and all; bread, stamped with date 1937, forty percent wood products; turnip soup, with an occasional sliced rutabaga; and potatoes, golf-ball size and smaller, dried or frozen and boiled in their heavy-hided jackets, perpetuated persistent flatulence. The regulars, with years of practice, had, first covertly and now overtly, fine-tuned orchestration. Depth of audio possibly relates to size of gut. The individual sphincter gives quality of pitch. The ensemble is predominantly after "lights out." As for me, after seven weeks of 20-20 ears and blinded 0-0 eyesight, it didn't make much difference, day or night. I remain covert.

With the casualties of the Battle of the Bulge arriving, and sounds of frozen toes dropping, and the introduction of dysentery taking its toll, the original ensemble broke up.

Other than our ground crews and squadron ops, I have been with few enlisted men, but the ones I do know are tops. Our American GIs in Stalag IV-B seem no different. The sharpest POWs, officer or non-com, pay attention to the British. Their systems function smoothly. Most have straightened out the bad eggs along the way, so the few weeks in the ward with Americans and British is a great social experience.

Jene and I as Fighter-Bomber pilots are as close to the combat troops as any airmen I know. We'd sweated targets that are in turn sweating our troops. Our relationship at Company-Squadron level is one of respect. Many have said we do our jobs better because of the others' sacrifices. We can see theirs and hear the concern in the Controller's speaking tone. Wiping out opposing tanks and artillery is our job.

Just a few years ago, I called a squadron mate in Maine. He came to the squadron as a Flight Officer, probably meant he'd walked a lot of tours. He was a whale of a combat pilot. His Boston accent put enough torque on certain words that he could be hard to understand at times. He and I shared several great ground-attack missions. We reminisce enough to get communication going. All the time we are talking, I can visualize his slight figure in the cockpit and remember his excellent eyesight and top notch air-to-ground accuracy.

I remember his thousand hour "crusher" hat and the fanged, lip-curled, wolf head painted on his leather jacket, shared with large wings radiating rays of glamour and gleam. "Boston Wolf" is the name on the jacket. "Lieutenant Davis" sobered a little but not a lot. His rank progressed. His skill leaped more. He related one particular mission handed to our squadron by Ground Control from General Grove of 6th Armored. German tanks are dug in, in woods. They and their infantry are nearly invisible. It would cost American lives to do it the ground way. Thunderbolts pummeled the forest-protected German troops by the acre with 50 caliber and rockets; three flights, four ships each. When Willie, Lt. Davis, got back to base there is a call from a General. They want to show the leader of that squadron just how effective the operation was. Early next morning, they sent a rig to pick Willie up. He is taken to the Division Command Headquarters, given a cup of coffee and introduced to three or four members.

The General burst through the door saying, "This, our fly boy?" Pounded Willie on the shoulder, spilling hot coffee on Willie's groin, burning him; the General never stops, said he had to shove off, tossed a thanks over his shoulder.

I am not sure of the General's identity. I didn't want to interrupt Willie's flow of words with a deaf, "What?"

Willie said the carnage was everywhere in the trees; the bodies just as they fell. It is quiet. His guide explained the destruction is not unusual, but the predicament they'd been in caused them to be profoundly thankful. Willie still has vivid visions of the deathly battle scene.

Willie wore coffee camouflage until he got back to Etain.

Nearly nightly, we hear shows by the RAF and daily looks at the working 8th Air Force bombers. We mostly hear their armadas and warning air raid sirens. We are not concerned a lot about being bombed, but we are told to strictly follow Jerry's rules and to refrain from cheering. That is an order. New American POWs are helped by the British.

I am managing pretty well in the ward. Jene's bed is taken by a Danish policeman. The Germans have imprisoned the entire Danish Police Force. He doesn't speak English nor I Danish, but he is a regular guy. He has a terrible goiter on his side, big as a hot water bottle. He doesn't complain but has trouble slicing or spreading his ration. I help him when I find one soldier charging him for help by demanding food as payment. The Dane receives food from home, sausage and cheese. He doesn't eat a lot and readily shares food with others. I can't believe my ears when I hear one American tell him he will cut his bread for a named item of food. This gnawed on me. I used a method passed down in my Mother's notes describing her aunt's story about a person taking a colt she adopted and nursed on the Oregon Trail. "I jawed him good."

We barter but care for each other. There is a difference between barter and sharing. We always share rations and the work or a need if we are able.

I continue carrying the briquettes and walking the perimeter of our compound. The German doctor comes and inspects us in our ward. My eyes are better, so are my legs.

I am saving a part of my ration. One of the Padres alerts me, we might move soon. Waves of bombers do day and night raids on Dresden, thirty miles to the southeast. Rumors circulate about the closeness of the

Russians. Our camp is on the Russian side of the Elbe.

I am told to get ready to go to the main camp. I will be shipped out tonight. I wear both sets of underwear; the British shirt, jacket, and trousers, my boots, and have one extra pair of socks in a jacket pocket. My wings are pinned to the jacket. Peggy's Testament is in my breast pocket. The shirt tail of my old shirt is used to wrap my book of "V for Victory" matches. I have the remaining cigarette lighter, my Russian aluminum spoon, and a table knife full of nicks. The knife saws wood and cuts tin cans as well. I still have both wrist watches, my insignia, dog tags, 9th AF patch; a GI can opener Jene has traded for with a couple cigarettes. My belongings fit in half a Red Cross parcel box.

Very valuable to me are the names and addresses of British and Yanks written in Peg's Testament.

Mid-afternoon, I walk with escort down to the main camp, probably half a mile. I am to see Jack Thatcher. I didn't see Jack Lomath except for the one time in the ops room six weeks ago. Jack T and his combine greet me warmly. Jene looks good. No bandage. My neck, forehead, and thigh are bandaged. Jack's combine has a late tea and supper. Jack gives me the "gin." "Dave, it's good to see you. You're looking good. Jene is wiry and doing fine. You have your own ration. We will supply tea. Do you have a cup?"

I show him a tin can recently acquired. He approves. Jack has scrounged one wool blanket. Says to me, "This is your blanket. Be decisive in how you say MY BLANKET."

I nod. He continues, "It won't be easy. When you leave the gate house, the guard will ask for your blanket. Hold it firmly and say, "IT IS MY BLANKET." Actually, it is Jene's and my blanket.

I eat a Canadian biscuit and a little cheese and part of my bread ration. Jack's combine's room is about twelve by fourteen with double bunks all around the edge and a crate-built table in the middle. Neat and looks to be the work of several years.

Some of these men built the stage set for the variety show in the fall of 1944. (Depicted in the TV film "Jenny's War." The setting for the book, "Jenny's War," is in Stalag IV-B's camp hospital in 1944-45, simultaneous to the time Jene and I are there though we are unaware. The TV movie places the time two years earlier than the book.)

Jack T is working at the main gate, his regular detail at the camp. He told me a couple of his stories on his visits to our farm in the sixties. I don't know if Jack is on the escape committee. He could be. In later years, he tells an eye witness shocking story of an incident in the camp office at the gate, Jack's duty station.

A POW work detail is assigned to do yard work. The German Sergeant is standing back from the window watching prisoners clean the weeds between the wooden walk and the side of the building. Jack watches dumbfounded as the guard aims his Lugar through the window. The worker has just picked a ripe strawberry and put it in his mouth. He is shot dead through the head at less than five feet.

Jack told me about the liberation. Mongolian Cossacks ride tough little ponies right through the open gate, clear through the camp, turning Ruskies loose and pointing them east, with nothing but the land for survival.

A day or two before liberation, a German officer, whom he had known since Africa, came to Jack, on duty in the gatehouse, saying, "Jack, we know you have a radio." (Jack says he is wrong on his count of one because we have two.) "The war is kaput. We'll leave soon. I would very much like to know how you have concealed it so well. Will you tell me?"

BBC news is distributed by word of mouth almost daily. Jack replies, "That's a reasonable request. I'll show you."

Jack takes him into the Gate Guard office, tilts the desk chair, slips open a false bottom and shows him the radio. Jack says, "Coo! He bloody well has been sitting on the receiver."

Jack's visits to the ranch end too soon. He died on the operating table in the early days of open heart surgery. Peg and I were weathered away trying to fly to Vancouver BC for his funeral. We turned back at Bellingham, WA. I hadn't been flying long after a thirty year layoff.

After the war, Jack had sailed all over the world as a bartender, purser, and entertainer aboard ship before marrying and settling in Vancouver.

When he planned to drive down for a visit, he'd give us a call. I can almost see him arrange his face and tongue for a "Western How do." It came out more like Reggie Van Gleason. He never fooled me, but he was much better at trying after I gave him a gray Stetson. He gave me his colorful little tam road cap. He drove a little Sunbeam. Our kids loved him and his car. His red-haired Irish Terry came, too, though just a couple times.

Jack becomes serious one day when we are driving a wheat truck to the elevator. It takes about an hour for a round trip, gives us talking time. I have driven the route nearly forty years, downhill, drops less than ten feet per mile, no load on the engine, quiet. Jack's story needs quiet.

The Russians plundered and pillaged. In Stalag IV-B, they had a separate compound. No one went there. Some would come out on work detail, and one came to the Lazarett, but they were given food that hogs would scorn.

Jack's story begins with one of the Padres asking Jack if he would round up a crew for a humanitarian mission. Jack found a team of horses and cart and three or four compatriots. The Mongolians and their Mongolian ponies galloping down main street of IV-B are welcomed, but the atrocities evident in Muhlberg's rape and looting clearly had given Jack a lifelong memory, an experience he wanted to talk about. He tells of hanging bodies, men and women, some by head and some by heels. His crew cut down, picked up and gathered corpses. He describes finding the Mayor, his wife and children's bodies, apparently shot while sitting in chairs in a family semi-circle. Jack interpreted the scene as the father shooting his wife and daughters to prevent their capture, then himself. He described the room in detail.

It's been a long time now since Jack's telling. I don't recall particular items, but I still feel his tone of sorrow and futility. The villagers of Muhlberg were caught as victims. Their town became terminus of years of thrust, siege, route and retreat. Now, it is a halting point, squeezed on a river separating two victorious armies. A third is in collapse.

Later, on the second day of their merciful voluntary task, they are driving along a tree-lined track when out of a roadside thicket steps a Wehrmacht Hauptman (Captain) confronting Jack who is walking beside the cart. Jack obeys his request, halts the cart. The officer indicates that Jack should follow him. In only a few steps the trail enters a house-sized opening. "I am looking at what seems to be a unit of forty or more German soldiers surrounded by a thicket. I says to meself, 'Coo and now I'm captured again'."

The Hauptman speaks "British" English. He tells Jack he has been observing what Jack and his crew have been doing for the village and offers his appreciation for their humanitarianism. He would like to surrender his command to Jack. Jack declines, saying he will report their predicament to the Senior British officer in camp.

Jene and I visited with Jack and his combine for awhile. Jack oversaw the creation of our pack, made from a reinforced Red Cross Parcel box. We combined our kit with room to spare. I am to carry the blanket, rolled with ends tied, the blanket over my shoulder sash fashion. Jene carries the cardboard Red Cross Parcel tied like a stadium vendor's hot dog tray, cord behind his neck.

Chapter 13:
Had a Beer

Train Ride:

Jack is on night duty at the main gate. Jene and I are to leave shortly after midnight. We will get the 'gin' from Jack then. We review our situation. We know all American officers who are able to stand are being evacuated. We don't know where we are going or how. Not much different from cadet days. Jene has been in the Main Camp about two weeks, long enough for him to have gained additional Kriege skills under Jack's tutorage. They are kindred souls, both bachelors and people wise. I knew more about cattle.

There are times for rest and times to go your best. We sleep first. We are nudged awake and escorted to the gatehouse. Seven of us are assembled in the building. We are a ragtag group of First and Second Lieutenants. Jene and I appear to be the only Air Force troops. We move along single file. Jack grasps us by the shoulders as we file out through a narrow door, nodding, "Good luck."

The door guard tugs at the blanket. I say it is mine. He responds pulling firmly. I hold on tight and repeat, "The blanket is mine." He turns loose.

It is dark and cold. Snow crunches under our feet. We walk single file with three guards, one in front, one in the rear, and one overseeing; an elderly sixty-plus Sergeant, a thirties Corporal, and a younger Private. The walk may have been two miles. A train stops; we board, occupying two compartments for the ten of us, seven of us POWs with everything we possess, most of it clothing on our backs. We are silent, each sizing up the others. Neither Jene nor I have met any one of our travel companions before.

Shortly after daylight, the Sergeant stands in front of Jene and me, requesting us to take off our pilot's wings and to put them out of sight in pockets. He indicates it will make life easier for him and more healthy for us. I don't know how he gets the message across to us, but he does very well.

One of our group speaks some German and is in much better condition than the rest of us. He is suspect. Maybe more recently captured or just better fit.

The train stops fairly often. About mid-morning, we travel slowly through a flattened city. We stop and unload on an open passenger platform. I cannot see anything standing. Much mangled track is heaped in places. Men are working in small groups. A train of gondola cars passes close to our platform. The open cars carry hand tools, wheelbarrows, hand-type track implements and men in striped pajama-like garments. Some have caps. All are gaunt, worn-looking, with listless expression. The appearance of forced laborers is shocking to my eyes. The gondolas have sides slightly above waist high as I look down inside them. They are moving a little faster than a fast walk. I think, "I can easily tumble into one." Reasoning quickly vetoes the idea.

This is the morning of February 19th, 1945. The British night raids of the thirteenth and fourteenth and the American raids of the fourteenth and fifteenth of February had devastated the city with fire storm bombing. We heard the planes and the bombs at IV-B. It is remarkable that trains can run. We are three days of bombing and five nights bombing later. One source says the two hundred plane American raid on February fifteenth is to disrupt the rail system to prevent retreating German escapees from the advancing Russians. Dresden slave laborers are and still are the startling revelation of the day.

It is clear why we, in attack missions, had to continuously, time after time, break rails. A significant number of our October and November missions were successfully executed imbedding delayed fused 500 lb sometimes 1000 lb bombs into railroad fills, destroying track and transport; but slave labor kept the trains rolling. You might say we and they worked on railroads, night and day.

The train stops often, however I feel reasonably safe riding passenger cars, even though it is in daylight, unless we are parked beside freight cars.

Jene and a couple others are plagued with the "GIs;" semi-thick, light brown, smells horrible. We contrive a make-shift cardboard receptacle to use. They are weak, desperate, and practical. I feel so sorry, but their adaptation in misery is to be admired. They could have just let go in the corner by the door.

We stop in a town within a narrow valley. Our train is one of several parked on parallel tracks with a covered passenger platform beside our car. Sirens signaled an air raid. We feel uneasy, but our guards indicate this is "old hat," probably bombers going to a more important target. This abruptly changed with smooth purring in-line engines, then fifty caliber machine gun bullets hammer in close. Guards, "GI" sufferers, and all bail for the deck. I forget stiff legs, feeble vision and bandages; hobbling I almost keep up with the Sergeant. He runs for the stairs leading down under the tracks. I crouch going down the stair well; can't help looking back, two P-51s strafe into view, their 50s hitting up track two or three cars just as a flak car drops its sides and lid. It is on the next track over, it blazes away with quad 20s or maybe two sets of twin 20s.

I shake my head bringing reality, it is not a "FLYING ACES" cover, but real. First instant cheer for on-track team. Apparently the flak car in a station is justified; they thought so. Is this the simple reasoning that gets wars started? In November, our group found plenty targets more vital than passenger cars. We didn't have to decide. The war is ending dirty. I'd experienced slave-labor forces earlier this day. Whip-saw changing emotion, how much? How long? Is a Hollywood director directing? Who wrote or even rewrote the script? Has it reached critical mass?

Under the track, walls are lined with benches on both sides of the tunnel. I sit with our Sergeant. I see no one else from our motley contingent. I feel fairly small here. I have experienced buzz bomb raids in London, rode the subways, saw the people. Here isn't a lot different.

These people seem less matter of fact, sit quietly, close and anxious. Almost directly across from me is a Hitler Youth wearing a tan and brown uniform. His rifle's butt plate rests on the floor, its barrel extends by his head. Splashes of red trim his uniform. I throw my thoughts into neutral. I feel just as conspicuous as he looks to me.

The Sergeant responds to someone's question. My uniform, healing burns--is it curiosity? They may have thought I am a tank soldier. I am glad my wings are in my breast pocket with Peggy's Testament.

The all-clear sounds. We join in the exit, climb the stairs. Near the top, Sergeant stops. Our rail car is close; two figures are coming from under the car. Need four more. Sergeant made a shrugging gesture of futility, like "How will I round them up; where did they go?" He has over estimated our physical strength and neglected to fully comprehend how starved we are. His message gets through to me. I point to our car. We'd left with a "get out of here quick" focus. I follow my pointing digit, saying, *"Essen"* (food).

He gleams. *"Ja! Ja"* (yes, yes) *Essen!"* Soon his cadre and mine return, one figure at a time.

We don't get out of the mountains. The weather deteriorates. In the late afternoon, we stop on a siding. Apparently we will be here awhile. Sergeant guides us into a small rail car on a side track. "Our" car has a table down the middle and benches along the walls, a stove and wood box on the opposite end from the door which is left open. The windows are covered by heavy wire grills. The space seems to be about the size of the rig for prison work crews which Cousin Ed and I passed on our bikes in my visits to Portland. We enter from the rear in single file up four or five steps between iron banisters.

Sergeant, through our German speaking POW, explains we will be here for some time. We will eat from our own rations. The corporal will bring a bucket of hot water (from a steam locomotive). Sergeant will be back soon. Probably checking trains and space available. We seven are standing, spaced around the table, concentrating on Kriege bread, slicing it thin. I share my table knife. We don't notice a troop train stopping across one track from us. Our preparations divert our attention from the surroundings. We are really amateurs.

A very loud antagonistic voice bursts from an obnoxious bullying self-important German Corporal. He pushes aside our Private guarding the stairway. The tirade is loud enough to draw further lowlifes in support. Three or four others from the troop train are on the stairs behind him. There is no doubt that his tone is threatening and in my mind, he is nearing inevitable out of control, spontaneous, self-enticed, combustion.

We are a pitiful lot to look at, not too clean, three or four of us fresh out of hospital, others recovering from effects of frozen feet and other trauma of Battle of the Bulge captures. Some barely out of boot camp and ninety day's practice soldiering. Evil one seems spurred on by his backup. His voice draws German viewers to windows of his troop train.

Checking the situation took no longer than a quick cockpit instrument scan. The stove is cold and close behind me; a three-inch, foot and half long stick of stove wood shows in my offense-prep scan, better than empty hands. It is in reach behind me. At my right shoulder, I sense Jene; he probably has something else in mind. Maybe a poker. Others are also frozen. No one moves. It looks like three or four of us are alert.

My eyes catch a fast move behind our aggressor. Sergeant split the backers as he charges the stairs. In a flicker, he stands before the bully. Sergeant's words cow Biggy into a boot-licking mode. I am astounded with his submissive departure. Sarge dominates. Sarge's firm admonition concludes with two words, "*--American Offi-ceers*."

Sarge says those words, as though we are Patton's Headquarters' staff. The hot water helps our spirits. We are beginning to feel we are under the command of a leader.

It is dark when we leave our mini-prison and board another passenger train, also with compartments. We'd had a little rest, stomachs, even bowels, settle down. It isn't long before we are sidetracked again. We unload into a small, about 14x16, rail-side section building with a raised sleeping platform about thirty inches high extending across the width of the building. It is behind a briquette fired stove. The wind is blowing loud around the corners. All of us and at least one of the guards are on this semi-snug bunk. It would have made a cozy ski hut.

In less than an hour, I know I am going to have to go out in the blizzard soon. Hot water and hospital habits are affecting. I nudge the guard telling him my problem. He hardly wakes but assures me by tone and word, *"Ja, ja. Javoh"* (to go out, get on with it).

I open the door. The double track is lighted by an electric globe under a reflector on a pole. The wind is howling so it swings the light. Thick snowflakes parallel the ground. The light doesn't show above my head more than a few feet. I step around the building corner, somewhat out of the wind. Nothing, not wind driven, is moving. Snow will quickly cover footprints. Unless the guard checks, I'd have at least an hour's start. No blanket, few matches, no food, and no help. This is the first time I'd thought of a serious attempt at escape. I am getting better fast but actually have a long way to go. It is only five or six steps back to the warmth and dry.

The war is not going to last forever. The Germans are collecting

American officers as potential hostages. We know the American lines are closing in. So far, we have had an easy trip. I go back in.

Our train connections are working. Sarge has done well. We get on a daylight train in the morning. In the afternoon we stop for a time, then get off in a busy station. Apparently, we are changing trains and have some time. Sergeant asks two of us if we would like to walk through the station with him. His invitation is only slightly surprising. Actually, his offer includes a glass of beer purchased at a counter. I don't have a taste for beer, however had enjoyed a couple glasses in England. He hands us each a small glass of what I judge to be light. I am surprised at its mild taste. The other prisoner interprets Sarge's remarks, saying he gave an apology for the watered down war-time beer. Odd that my first glass of beer on the Continent is a treat from an enemy soldier. We have time to fix an early evening meal from our meager pack. Things are looking up.

We board a train after dark, settling into compartments much better than anything we have ridden before on this mini-sojourn. In early evening, Sergeant asks me and the afternoon second party to go ahead a couple cars with him to get information.

We enter a well-lighted, chrome and mahogany atmosphere of both men and women in business clothes, afternoon dresses, and coats and ties. A congenial gentleman advises us in excellent English that we will be getting off soon on the outskirts of Nurnberg, at a place called Furth. We will have to walk across the city to the prison camp. The allies, British and Americans, have bombed the railroad so severely the train can't get into town. The British bombed last night, the Americans today, and it is likely the British would give it another go tonight. He further instructs us, this camp is much nearer to Switzerland than we have been; therefore Red Cross Parcels and medicine are distributed much better. In fact, rations will be better here. Quarters are warmer and showers are weekly. "Blankets and beds are enough for everyone."

I wanted to believe some of this, but we took a wait and see attitude. We relate the story briefly before the train stops.

The station is small with faint lighting. The white and black sign under the roof says, Furth.

Chapter 14:
Steps and Sugar Cubes

We unload onto an ordinary suburban railway station's platform. It is concrete under an open shelter. A shaded light reveals a smoked white sign "FURTH." It is four or five steps down to ground level. We line out single file, same as we left Stalag IV-B only instead of crunching white snow, we head into flickering dark. Fires burn, the crumbled city lies ahead, smoldering, in silhouette. Three guards, seven American officer prisoners begin. We move unhurried but steady.

I judge we are starting between eight and nine o'clock. The only sound is the occasional "whirr" of the guards' hand-held, plunger-driven dynamo flashlights. I couldn't figure them out in Kaiserslaughtern or the first train ride. They have to be seen to understand them. They are better than a firefly. The whirring is as close to audio light as the hiss of a gas lantern.

We trudge along, no one speaks. I soon glean it is going to be a long walk. Air raid sirens begin wailing after about an hour. By now, we are stepping over fire hoses, parts of buildings are falling from today's bombing. Debris blocks our path. The little whirring lights are used sparingly. We don't stop. Tonight's target is close ahead of us. We hear people in agony.

I am glad I have been walking the camp's perimeter. I am probably about third in condition of the seven of us. The guards are patient, don't hurry us, but nothing seems to veer us from the steady cadence. The "all clear" sounds. Fires are burning, ambulances sound to be navigating. My interior clock is uncertain. Nurnberg might be a longer mission than Dresden last week; however, then the bombs were hitting thirty miles away; tonight they are falling from overhead.

Another raid approaches. Bombs drop, closer this time. The ground shakes. I judge we are in the heart of town. Sergeant stops right by a fairly rugged stone arch, just tall enough and space enough for us all to get in and under. His message is short and clear. The gist translates to, "If we do not get out of the city before daylight, I cannot be responsible for your lives!" His implied message is, "It is now half-way to daylight. You men pace yourselves." My guess is we also have half-way yet to go.

I think of Slim, a World War I American soldier on my first harvest job. His favorite descriptive physical phrase, from "GI 1918," is "Dain't dis de dose dof de dhits!" All words in his most dire description began with D.

Anti-aircraft batteries rain shrapnel. More bombs, more sirens, more fire hose, more rubble. We do a steady pace; more steps, no complaints, just step…step around, step over, step on, must keep stepping like piano teacher, Mrs. Irby's metronome. Step, step, step. Ignore bombs, careful, no stumbling, no slowing. We aren't hot or cold, just dog tired. Eventually it is quiet; our steps keep on. The last bit of shrapnel is down. We are moving over less rubble; then no rubble.

Fog settles around us, gives us extra time before daylight. Another twenty minutes, a half mile further we see shrouded water, maybe a lake, on our right.

We are through town. The street is wide. The fog swirls over a wall on our right. Two concrete steps stretch alongside. Sergeant leads us up onto a broad cement walkway. We are numb, yet dry, inside and out. We have made it. We are out of the city.

Sergeant says the camp gates are a half mile away. He also indicates it will be better if we arrive an hour later; now we should sit and rest. It is even difficult to get down. We settle on the top step in a row of sorts. Jene and I fix some kind of food, but mainly rest.

The fog begins to lift. We can see across the broad street in front of us. I remember a broad streets picture in my "CURRENT EVENTS" weekly paper in Olex Grade School. It is called the Autobahn and passes by the entrance to a stadium. The fog is lifting and also from my memory. I turn half around. Grade school recall cues a visual search. I look up. It is there! The huge Swastika. The hairs on the back of my neck rise. It strikes hard, hard like a flail hits a threshing floor again and again. Using a British phrase, I say to 'meself', "Coo, this is where it all started."

Steps and Sugar Cubes

My seventh grade weekly paper's story played up the magnificent achievements in Germany; new highways, new buildings--the Stadium!

We sit quiet, each resting. The long expanse of concrete wall extends what seems to be a football field's length in both directions. The vision of this structure, like the image in the mirror held in my hand in the dressing room at Stalag IV-B, two months ago, is a part of me for the rest of my life.

We Kriegies get up first. One by one, we are on our feet, all except one. No one utters a word of complaint. One of our seven, an uncommunicative Second Lieutenant opens his cardboard pack, takes out a box of sugar cubes and moves to each of us, inviting everyone to take two cubes, insisting this is what he is going to do. I am not sure I can melt one, but I do two. His gesture works wonders for us.

The Infantry Lieutenant with frozen toes is still down on the step. He tries to get up. Someone, maybe Sergeant, gives him a hand. I don't believe I see him make it to his feet. Jene and I understand each other. Jene picks up our half Red Cross box. I sling the rolled blanket over my head; only the cord is over my chest. Sergeant is looking at the Lieutenant.

No one says a word. The Lieutenant is fixed in place. He cannot move a limb. Both he and Sergeant seem to search to break the impasse. Sergeant had said camp is a kilometer farther to go when we sat down. He has given us a rest but also is gauging the time to arrive at the gate.

I grab my right wrist with my left hand, step in front of the man most likely of our group able to help and reach for his wrist, saying, "Let's carry him."

My cohort volunteer knows the "fireman's carry" probably from Scouts. I learned it from cousin Ed. My captors used it on me. We pick the Lieutenant up. He is light. Our slightly-sidewise gait seems to fit with my hippity-hitch stride. We move out together. In fifty or sixty steps, the youngster starts wiggling, says, "I can do it."

We are cheered and thankful. I couldn't have gone much farther. His grit and will boosted every ones' outlook. It doesn't take a lot to be an improvement.

We make it up the straightaway, turn the corner, move to the right along the fence to Stalag XIII-D's Main Gate. We see barbwire and guard towers but no action. We hold up at the gate. Sergeant eventually is passed in to gather the gin.

We cannot see anyone walking in the fenced compounds, which look to be large according to comments from those around me. There are no indications of the amenities offered by the passenger in the luxuriant train car last night, before Furth. He is correct about our night of probable bombing and walk. A camp like he described would show energy. We are rightly apprehensive.

Sergeant is let back through the gate to us, his stretched and sagging charges. We are let in the gates, and to the outside of a receiving building. It has a board porch, with overhanging stoop held up by 4x4 posts. He begins by saying we can sit down, rest. No one moves. He adds it will be about a half hour before anyone can sort us and assign us to compounds.

Waiting time of half hour courses through my brain. My question, and seemingly the one everyone is asking himself, is simple. "If I sit or kneel or anything but stand, will I be able to start again?"

Each individual votes with his body. We all, even the man with raw feet, stand or lean in place. I don't believe the vote would be different if the length of wait is an hour. The distance from chin to the floor is about five feet and from the floor back to standing, another five feet. Atmospheric pressure is around 14.7 lbs per square inch, at sea level; a falling object accelerates at 32 ft per second per second. Nothing figures. We are exhausted. Doesn't make any difference whether you apply a slide rule or a bull whip, we seven can only maintain until we are given rest and time to recover. Sarge understands. He undoes his puzzled look and speaks to his two assistants, then leaves.

Early one morning, decades later, a rare farm event illustrates fatigue and recovery. A flight of migrating swans land on our small pond along the private road crossing our field between the highway and our farm buildings. I drive three miles around 'the block' and put a cable and sign across the entrance to our farm road, stopping any disturbing traffic for a day. I watch the swans carefully with glasses and telephoto lens, several times. They appear to be a pair with five this year's youngsters.

The sun is setting on the Cascades when Peggy and I make one last look. If they fed, it is before we discovered them. The sun's last ray seems

Steps and Sugar Cubes 95

to be the signal. They lift off, seven at once, low and slow, like heavily loaded Thunderbolts, using ground effect. They circle once just above the water then rise a bit and pass over us where we stand. Gaining altitude, they line out, silhouetted against the glowing sky, flying south to southwest. The pond gave us many beautiful moments. This may have been the best.

Perhaps we also have the ability to go until it's all drained out of us. The unit, too, is important. Sergeant, the boy with frozen feet; we all contributed as each could.

We get our assignments. Jene and I are with a barracks of recently moved Air Force POWs. We go through a couple gates and to the barracks. We are instantly aware showers are seldom, if ever. Food is sparse, bunks are filled. Blankets are as scarce as food. Morale seems pretty good. Our mates walked and rode box cars from Poland, arriving ten days before us. Unlike us, all did not make it.

We are amazed at their talents and fortitude. They are Aussies, Canadians, Texans, Washingtonians and Oregonians, and a couple Montanans.

Chapter 15:
February 21 1945

My diary was carried in my head until the bandages were removed from my eyes on January 4th 1945, Ron Bleecker visited from main camp with Padre aided permission before he was sent out to a work party. This was our first time of actually seeing each other. He confirmed earlier dates. I entered short one line notes in my (Peggy's) testament.

Peggy's letters—after I am MIA

I continued writing letters to David every day after the MIA telegram, putting them away in a box. Letters filled with love and longing, past memories and future hopes, also any scrap of pertinent war news.

December 4 1944

...It came today, the telegram saying you are missing in action over Germany.... I don't know where you are. I don't know if you are hurt, but I know you are alive. I can tell. I know.

December 6 1944

I am so certain you are all right, not only for spiritual reasons, but for physical ones. You are such a strong person. Your body can take much punishment, you are well trained on what to do in any emergency.

I read an article in Flying about treatment of prisoners in Germany and if you are one, I should hear in February, about the 23rd.

December 7 1944

Pearl Harbor day! ...I pinned silver bars on your summer blouse today I borrowed a couple of Daddy's [Home Guard]

December 12 1944

February 21 1945

...And so another day goes by. Mom and I went to Grandma's today. I looked thru her old papers and found the Nov. 18 Oregonian and what happened in the air Nov. 17. It said that all bombers were grounded but hundreds of fighter-bombers bombed...six miles from our troops. The weather was cloudy and rainy.

December 17 1944

Wherever you are, darling, it's Dec. 18 and I know you are remembering our {wedding} day...The news is not good tonight. The 9th Air Force <u>fought</u> some savage battles and lost 35 planes.

December 19 1944

They say the Thunderbolts are flying in weather no plane should be in. (The Battle of the Bulge)....I read your letters over and over...

December 20 1944

I received a letter from the war dept. today about you. It was a regular form letter giving no new information, just that I'd get any details which they may learn, and in time receive your personal effects.

December 21 1944

...We don't know what is happening over there, except it is a terrible battle and many men are being killed... A F men are to carry weapons at all times for enemy paratroopers.

December 26 1944

The Chanel (perfume) came today...Daddy and I were home alone when the mail came. He brought your package and said, "Now, if it makes you cry, go right ahead."—but I don't cry.

January 2 1945

My Dearest One, Wherever you are--over there its January 3rd and you are <u>22 years old</u>...I wish I could give you a chocolate birthday cake...was thinking of two years ago when I made that awful cake with the pancake flour--and last year when I brought the cake out to the field

(Luke) and scared you to death with the knife I brought onto the post. heh heh.

January 13 1945
The Russian offensive has started. Who knows, perhaps they will overrun a Stalag Luft, and you, if you are a prisoner, will be freed--

January 16 1945
Daddy went to the Red Cross today. They cannot do anything until we know if you are a prisoner, but they said for me to write to your CO as they cannot. I wrote a letter to him tonight

January 19 1945
I had a letter from the Red Cross today...They say 65% of flying personnel who are missing are known to become prisoners...The Russians are at the Silesian border. According to what I have read, Stalag Lufts 3, 4, and 7 are not very far inside

January 22 1945
I rec'd a lovely letter from your father today. Honey, he is very splendid about everything and feels just as I do, that you are alive, and will come home.

January 26 1945
10 weeks ...since you flew away and didn't come home. But life goes on here. Daddy was gone this evening [Home Guard]...so Mom and I milked...Your wife is a whiz with the milking machine.

January 29 1945
Mom and I went to the blood donation center today and each gave a pint of blood. It didn't hurt a bit and gave me such a good feeling to do it, even tho it is such a small thing to do

January 30 1945
Just 8 months ago today, that we kissed each other and said "I'll be seeing you."

February 1 1945

Today 11 of the letters I wrote to you came back, all stamped "missing" by Capt--6 of them were written in Oct, one on Oct 2...It made me feel very sad that you didn't get my letter telling you how pleased I was with the lovely scarf--and most of all, you didn't have the one about your air victory. I wanted you to have that one very much...

February 2 1945

The news is good...but, today, they say, the allied prisoners, two camps of them are being moved in Germany, walking thru the snow...

February 3 1945

...David, I believe I will hear from you sometime between February 18 and 23...

...Then I say, Peg, that's foolish to fear, for what happened November 17 is long past and done with, and you know David is alive. Yes, I know – I guess this one little fear is the only weak spot in my armor.

February 10 1945

Quite an article in the "Oregonian" magazine section about Oregon's aces. Talks of Capt. Fisher and Mogin's Maulers--the famous $9^{th}AF$ P-47 Group. Dearest, how I wish you were with them [still]. Childsy, you did a great job, just as I knew you would, before whatever happened, happened.

February 12 1945

Mom and I got to Arlington at 8:30 PM. It is now 2:30AM. Your father and I have talked over everything.

February 14 1945

Mother and I left Arlington this morning. Your father is well, and is getting along fine. He goes out quite a lot, to church in town, has dinner there to games, Grange etc. He seems to have all faith that you are all right.

February 16 1945

Next week, I think I may hear from you. Next to me I guess Daddy is most eager to hear from you. He can hardly wait to hear what is in every letter I get, thinking it might be something about you. He has told mother that nothing ever affected him so much as you're being missing.

February 17 1945

Today the Russians captured Sagan, where Stalag Lufts 3 and 4 are--or were. They don't know yet if the prisoners were transferred.

February 20, 1945

Today a letter from the war dept came--

Excerpt from letter from Headquarters, Army Air Forces to Mrs. Peggy S Childs; received February 20, 1945:

...with reference to your husband, First Lieutenant David A. Childs, who has been reported by The Adjutant General as missing in action over Germany since 17 November 1944.

Additional information has been received indicating that Lieutenant Childs was the pilot of a P-47 (Thunderbolt) fighter plane which departed from France on a reconnaissance mission to Saarbrucken, Germany on 17 November 1944. The report reveals that during this mission at about 11:00 a.m., four miles northwest of Lambrecht, Germany, your husband's (fighter) bomber disappeared from the formation while strafing enemy locomotives. It is regretted that the circumstances relative to the loss of this aircraft are presently unknown in this headquarters...

This letter is informative but not too reassuring.

David, I wish I knew what this means...flak...engine trouble...too low to bail out...crash landing...prisoner...I'm glad to know these few facts tho...

February 21 1945

It is just another day, about ten o'clock in the morning. I am in the living room and happen to glance out the window. A car is speeding up

February 21 1945

the road past the mailbox, coming towards the house. A blue car. It looks like the mailman's car but its two hours until mail time and anyway it didn't stop at the mailbox. The car rounds the corner below the house. It is Ellis, the mailman. Something explodes inside me, and I run outside and meet him as he opens the car door and hands me a card. Oh, precious card! He's alive. David is alive!

Down at the barn I see Daddy and Jack come out the door to see who drove in. I wave the card and scream the news. "It's from Dave! It's from Dave!"

The two of them leap into the air jumping and shouting, looking like two animated stick figures. Bless Ellis. He and the Postmistress have shepherded David's and my letters since the first ones the summer of 42. This special trip with this wonderful news is an incredible gift.

The Card--

Kriegsgefangenenlager   Datum 1 – 13 – 45
Sweetheart, I see no other course
but to face it. I'm living for the future,
but am doing O.K. in the present.
I love you like you done never been loved
before, little darling. Cigarettes are very
few and valuable, now days. You might
send a thousand, with lipstick-tip preferred.
Dave

(This first card which Peggy received was written six weeks after my first two sent December 2nd and 3rd. This January 13th card was in my hand printing. The first letters contained a note saying I had a minor hand injury and were written for me, though I was able to, unseeing, sign them. The first sentence, "I see no other course but to face it." was an in-house code line we had acquired from a radio drama for any situation not in our control.)

February 21 1945
…*Today the card from you came. Ellis made a special trip with it. David, I've been in a rosy haze all day…I thank God that you are alive. I*

don't know what you've been through, where you are, or what you have yet to face...but you are alive...

Remember I said I'd hear from you between the 18th and 23rd? Your father counted on the 23rd as his lucky day.

February 22 1945

...so many people have told me today, how glad they are that I've heard from you. Everyone calls you Dave, some have tears in their eyes...The word got around that you want cigarettes. They are very scarce. People wait in line for them. But I'm going to be swamped by them before I am able to send a package to you. I have to send a Photostat copy of your card to the Provost Marshall's office in Washington before I can get a proper label to send a package.

Chapter 16:
Nurnberg

We are assigned floor space just inside the front door in the left corner. Pretty good as floor space goes. There is room for most in bunks. Most also have two blankets each. You would think one blanket between two would leave us only a half blanket each, one fourth the normal two blankets. Not so. Know-how counts. It was given to us straight and practical. One blanket, two men, board floor, no bed. You simply place:

1. Blanket on the floor.

2. Lay down head to foot, fully clothed, side by side, about one turn from the end of blanket.

3. Wrap arms around each other's ankles, shoes on feet.

4. Roll up in blanket.

The variables and adjustments soon iron out.

Jene's main camp experience gained benefits. Our silver bullet is the can of DDT he traded for with one of the newly arrived American POWs from the Battle of the Bulge. First dust the blanket, then our clothes. We are lice free for the duration. DDT is new to us. We dust once a week, like we treated hen houses at home.

We never figured the calories gained by not feeding the little vermin. We are close lipped about our war experiences, but we exclude even any thoughts of the valuable dust.

New to me are several things like *"appell"* or roll call. Every morning, rain or shine, we line up in ranks five deep. Guards walk down front rows, saying, *"Eins, zwei, drei, vier, funf,"* counting each row. Ranks are five deep with the counting guard passing in front of us and the checking guard behind.

If the count is incorrect, we stay out until it is correct. If the numbers are incorrect and someone is actually missing, we stay out while they search the barracks. They take their time. We are denied a bread ration for the day.

The compound has a sparse library; prisoners give lectures; discussion groups meet on almost any subject, many spring up. I try to learn bridge. Problems develop. I can't see across the table, my eyes goop up. Jene is an excellent player, teaches me enough to enjoy listening to and "watch" his bridge.

Water comes from a single stand pipe, two barracks up. The camp is near the railroad marshalling yard. British Mosquito bombers deliver 2000 lb. bombs to the tracks almost nightly. Mosquito engines are loud as they circle low and precisely orient for bomb runs. German flak fragments fall on our roofs. We are not allowed out of the barracks during raids either day or night. Neither are we allowed to enter the camp's idle bomb shelters.

I go to sick bay for bandaging weekly. We arrived on February 20th. My final bandage is March 6th, one-hundred, thirty-four days since I was downed. The fellows in the aid room are cheery.

Rations are slim; bread is divided between combines according to number in the combine. Food is divided by the honor system along with safeguards. Safeguards are: The man who divides, slices, pours, or counts the ration gets last choice: Another way is cut the cards for choice; a third method is--one person points at a portion while a second person, with his eyes turned away, names the recipient.

Sections of the barrack's one-hundred-sixty men line up for the daily soup distribution. Food becomes the total focus. Our GI can opener, a trade from 106th Division POWs in IV-B, and my table knife are continually in use. They are always returned. The Germans are giving us potato soup with a little barley sprinkled over it. Soup arrives in a tub carried by two Kriegies carrying the large open container with handles. Each of us has some sort of cup, dish or tin can. Size only need be large enough to hold the portion from the dipper. Jene's and mine are tin cans brought from IV-B. It takes confidence, skill, and a Solomon to distribute the soup. We line up, all of us, inside the building with a kind of honor system for "seconds." There always are seconds to insure everyone gets equal distribution once. Seconds rotate. If you are among first today, you should be near last tomorrow. When seconds run out before you, you get first for seconds next day.

The luck of the day places me right behind the next man to be served in the queue for "soup seconds" when the pot runs dry. Instead of us being first for tomorrow's soup ration this innovative person, "Solomon" holding the ladle looks into his empty pot, but for three non-liquid bits and pieces. Choice bit, number one, is a chunk of meat about the size of a 1940's large Baby Ruth candy bar. Item number two is a combination, a chunk of meat same shape but only one-half as long as the first, but it is attached firmly to a femur bone from a large critter, knobs and all, about twenty inches, an oxen's? Maybe it's an artillery horse, a Thunderbolt road casualty. The third item is an egg size potato. These three items are all that is left in the big pot. Solomon gives the enabling words that turn away those behind us. "First three have first, second, and last choice."

First man chooses and receives the largest bit of meat in his cup and starts away. I reach for my choice, already thinking how we will each relish a bite of meat and figuring how to get the marrow from this monster bone when First Chooser utters a near growl, "I got first choice. I get the bone."

He throws the first meat chunk back and grabs the bone with its attachment. I look at Solomon. He is perplexed; neither Solomon nor I fathom this turn. Division, choice, shares, are by custom, accepted in trust. This extending the first choice to first choice twice shakes me. The division was decided. I am satisfied either way. In fact, pleased with our second, second choice. I get an unspoken message from Solomon. I was on the verge of snarling back with an animal retort. I didn't. Disgusted, I look squarely at Double-First and say, "You Shit." I am shocked with his act and my react. I didn't know him or recognize him again.

The hot water with ersatz coffee mornings is no problem. Seems to be plenty. Jene and I take turns slicing the bread and spreading jam or margarine on bread. We always have something, but it is only because we always ration. The British taught us well.

Folks practice cutting their bread ration as thin as they can. In the toughest of times, we get one Kriege loaf "between" ten men per day. This means each of us gets less than an inch slice of bread. We can generally slice it into eight thin slices and not lose a crumb. We save the slicing dust as well.

We gradually succumb to less and less interests as the food becomes shorter. Eventually, only the super savers have supplemental food. Red

Cross Parcels become non-existent two weeks after the last one was divided, one among twelve. Men educated along these lines determine our total ration is less than six hundred calories a day. I know our soup is probably less than two hundred. I can't figure more than three hundred calories a day. I see three potatoes from thumb size to two thumb size. Soup is the water they are boiled in. Those who ration by themselves or bash their reserve are in dire shape.

The hunger is strong enough that not much else is on our minds. Not complaining hunger--just Hunger. Each combine severely rations food. We all are hungry. All want to survive. The fate of prisoners is questionable. We are organized; all plan to survive.

A supposedly newly captured American is brought to our barracks mid-morning. The "new" POW smokes his cigarette really short. Old Kriegies take a few puffs then put them out and light later for a second round. Finally, before they are as short as the new guy's, a careful smoker puts the butt in an empty pack with other short cigarette butts and eventually uses them to roll his own. This fellow <u>steps</u> on his smoke. He probably is a newcomer. Doubts ease somewhat. However, he may be a well trained imposter.

We are interested in news. The story is Germans brought him almost directly into camp, downed by "ack ack" on a daylight raid someplace near, like Stuttgart. Prisoner interrogation has apparently become a casualty because of the losing of the war. This is a sign of better things to come but neither I, nor anyone else, digest this information. He sits in a circle of about eight on chairs and a bench with open space in the center. I am on his right, one person between him and me. I hear very little he tells us. Perhaps no one else does either. He is hungry for he seems to be relishing a part of his day's ration of German POW bread. What holds my attention is, unknown by him, a crumb of his bread drops and rolls across the floor and stops beside a man almost opposite of him. It is small by new guy's standards if he even knows. He doesn't know that it is huge by our standards. It lays about a foot from its nearest person.

No one listens. He talks on. I glance at the others. No one focuses on the crumb. It is scarcely half the size of an ordinary pencil eraser. Everyone knows. Everyone sees it when it sort of bounces once on the way at a rough board. All look away. We ourselves are the diversion. The new man discovers he has lost his audience and gets up. The circle shifts. The closest man, in one smooth motion, picks up the crumb, stands up, turns and rubs his nearly closed lips. His jaw never moves. No one's eyes meet. The crumb is divided correctly.

It is cold nights. Jene and I never get bunks. We see and learn of the Kriegie burners brought from Poland. Some are made like squirrel cage blowers fanning a tin fire box, all manufactured with tin reclaimed from Red Cross food parcel tin cans. The burners are each unique, hand crank driven with carved wooden pulleys and braided rope fan belts. There are also simple burners made by fastening two Klim (milk spelled backwards) powdered milk tins tied together with punched holes and wire fashioned into pig rings, open end to open end, with the bottom one formed with split out legs, and inside supports holding an internal smaller can for a fire box. The fire box has side holes that ignite when burning wood gasses emit. The idea is similar to an early oil burning heating stove. The fire box is about two and a half inches in diameter and three inches high with a split door about an inch by one and a half inches. Fire wood is shaved from fence posts, small pieces of siding or a broken bed slat.

The little Kriegie burner is efficient, produces more BTUs per cubic centimeter of fuel. It takes a lot of support work. Water boils in seven or eight minutes for a pint and a half. The wood is cut about the size of a kitchen match. We hand feed it, one stick at a time. We built the stove using our can opener and table knife and searching in the can pile. A golf tee would be a good size for a back log.

On March 18th, I write *"Most food yet!"* We receive a Red Cross Parcel among eight people. This means we two, Jene and I, receive a fourth of a parcel, two and a half pounds. We take some Klim, oatmeal and part of a can of sweetened condensed milk, add water and mix it in a can the size of a two pound coffee tin. We stoke and stir for maybe thirty minutes. We get nearly a pint each. WOW! It is good.

We are one of many groups scattered around the *appell* or roll call grounds. The General Staff of the Serbian Army is set up cooking nearby. Several Serbian officers, also POWs, watch as we concoct breakfast. In reality, it is also lunch and dinner in a can. I josh Jene with, "No chef has been studied more closely." However, I discourage him from taking a bow.

Our audience is conspicuous because of their once fine uniforms;

broad lapels of color; some wearing riding boots and magnificent tailored great coats. Jene and I adjourn to the barracks before dining. Next day, we try toasting cheese on Kriegie bread.

We can sense minute changes for the better. Back in February, soon after Jene's and my arrival, a little before we got reduced to a "stone soup" ration, an enterprising soul confides in us, asking for a couple lumps of sugar and a few raisins. He wishes to create a bottle of wine to toast our liberation. We buy in on his patriotic stealth project. The bottle is "aging" for five weeks. Our sips are minute. It is fortunate we have plenty of shareholders for only one round.

I can say that it isn't the best or only wine we taste in Kriegie camp. Somehow, the Padre prepared Lutheran Communion on Easter Sunday, April 1, 1945. I think of Chris and his and my Communion in the "chapel" barn in Brittany eight months before.

A week later, we receive Red Cross Parcels, but also spread thin. We are given squares of wax paper and a recipe for building emergency rations. The ingredients come mostly from our shared supplies. Everyone in our compound goes to work with crackers, D bars (a kind of an American durable chocolate bar), some kind of dried fruit, jam, oatmeal, sugar, dry milk, margarine and crumbled bread parts. The purpose is to make a three day ration, two bars per day, loaded with energy. The process is demonstrated in each barracks. We mix and knead the ingredients and form heavy fruity bars. Each bar is securely wrapped in wax paper. We are to carry the bars on our person at all times in case of any disruption.

We are now into better weather. Some sunshine, along with a little extra food, helps our outlook. My notes in Peg's Testament for March are: *March 6th, final bandage; March 15th, first shower; March 18th, most food yet.*

The shower is conventional. (Similar to gas rooms in death camps. We are not yet aware of the death camps.) We receive water for two minutes under the ceiling type shower heads. Two people per station makes it one minute each. We alternate ducking under and out, about three rotations. The water falls directly down.

The shower is the first time my face is exposed to splashing or for me to know I am not able to close my eyes at all. The water stings the outside edges before I can turn away. I long for a good old back porch, harvest time, wash pan, two-handed face scrub with streams of water running off.

In conjunction with showering and retrieving our clothes from delousing, we have time and opportunity to see a scale and to weigh in British units; Jene, 7 stone; I, 9 stone. One stone =14 US lbs. We are 98 and 126 pounds. My squadron weight was 175, Jene's about 145.

April 1st: Easter Sunday. I write a commentary of three lines:
P-47s; V for Victory in Morse code of clouds; Communion, Lutheran Chaplain; lovely breakfast. (Another oatmeal bash.)

A flight of P-47s flies over the camp. At least one barrack's roof is marked with a Red Cross. A group of clouds drift over camp in the ...-V for Victory Morse code configuration. Communion is the same as in the Brittany barn. Our group Chaplain and the Chaplain at XIII-D are both Lutheran. I am country Methodist Sunday School with a preacher visiting occasionally. Dad is Methodist and Mom a Baptist.

April starts with anticipation. We are, however, apprehensive by the morning of April 2nd. Camp is moving. One group will travel by boxcar, the other will go on foot. German doctors look us over. Maybe we are pointed out by POW doctors. Anyway, Jene and I want to be in the walking group heading for the next camp. We are told, "About one hundred-ten miles south to Moosburg, near Munich."

The one hundred miles seems to be known. They are moving us closer to Red Cross Parcels coming from Switzerland, yet farther from Allied Forces. The move also seems to add credence to rumor, Hitler wanting to use POWs for bargaining chips in one of his schemes. Adding to problems of either mode of travel is chance of "friendly fire." Both methods of travel pose a threat from ground attack strafing by the inexperienced or over eager in any allied category. We prefer freedom to dive into roadside brush over being crammed into a box car, "fish in a barrel" like exposure.

We both attempt to qualify for the hikers' contingent of several thousand POWs. A German doctor decides who will hike and who will hitch. We are under scrutiny to walk down and back a few yards. My left leg won't straighten. The dip in my walk relegates me to the group of two thousand riding in wretched confinement in the box cars. Jene flunks the physical test also. Our successful combine keeps intact.

Next morning, we move out of camp to the nearby railroad yard. We wonder if we will see the walkers again. We line up in groups the size to squeeze into a box car. It is up to us to paint the letters POW on each car. A tub of thick whitewash and a broom-brush passes from car to car. POW is painted on both sides of every car. I am in better shape than most and climb the iron ladder to the top of our car. The roof is more oval than the roofs of cars we load wheat in at home. Engineering drawing class back at OSU is brought to my first practical opportunity; in fact you might say it can be a life or death matter. Men standing on the berm alongside the track dip the brush and hand it up. I apply the brush and paint in wide block letters, POW. The letters are bigger than the "bull's eye" on air to ground-gunnery targets in flight training. I don't want to ride the train, but being as we are, we make the best of it.

Chapter 17:
Water Bottle

The train ride is made less miserable because of the experience senior officers gained from negotiating skills learned while leading our mates from Sagan Luft 3 in a snowstorm, in the middle of the night, with the thermometer at zero degrees Fahrenheit. It was accompanied with sounds of, not distant, Russian artillery.

Some of the guards on this train were with that February exodus. Jene and I sense we are getting beneficially pre-flighted by our mates in their trek two weeks before we arrived at Stalag XIII-D. We also know rations here are less than one-third what rations were at Luft 3. Another big difference, now it is warm early April instead of an all night blizzard trek in driving snow.

My train experiences thus far, consist of a German hospital train, followed by my cold boxcar on-a-stretcher ride the last of November to Stalag IV-B, Muhlberg; followed by the ride of we seven POWs from Stalag IV-B to Furth in passenger car compartments, after which we trekked on foot to XIII-D, arriving exhausted, only shortly out of the infirmary in Stalag IV-B. Now both of us have lost fifty pounds since bailing out, yet we are somewhat stronger than when we arrived in Nurnberg.

We start boarding the box cars. There is some straw on the floor and a half-barrel latrine in the center. These are extra niceties planned for our comfort.

We fit in with the seasoned POWs from the Luft 3 exit march. It is not difficult to envision that truly we are gaining from their experience. Straw and latrine half-barrels are big improvements over the cattle dung laden, frigid express they traveled in coming to Nurnberg.

We sift into every space. The mass of sixty men is now able to fit in where two months before, fifty were extremely crowded. Reason enough, sixty probably now weigh the same as fifty did earlier. We also are not carrying nearly as many Red Cross supplies. Then Kriegies loaded knap sacks made from shirts, and suit cases, cleverly home-built from Klim tins or wooden crates, loaded with a departure distribution and or help yourself initiative, from camp stocks. Much ended in snow drifts along the road in their miserable days before boarding wretched and fouled rail cars.

Finally, we have learned to get better cooperation and attitude because of eminent defeat for our captors. I'm sure the Kommandant and our senior allied officer discussed and ironed out problems that led to near revolt in the early February situation. Only the less able POWs are on board the train. It is possible the guards, too, are less fit than those escorting the marching group.

I didn't recognize the subtle benefits coming our way. It isn't that I am unappreciative. In fact, four years before, I asked my father to let me ship a rail carload of steers to Portland Union Stock Yard's market in hopes I could ride down the Columbia River Gorge in the caboose. We shipped them by truck. I didn't ride either truck or railcar; didn't miss days in school. I'd get plenty of experience later.

There isn't slack room anywhere. We sit on the floor, knees drawn up; only about a third can lay down until we settle in tired. We manage some sort of water containers to get tea along the way. I feel good having the big white POW letters on every other top and both sides of each car. They can easily be seen from a distance.

There is not a lot of tension in the car. There is some justified anticipation of improved food and facilities. It is deep dusk in the car. Slivers of light squeeze through very narrow cracks. The car has probably hauled bulk grain. It has wainscot two-thirds of the way up on the ends and walls. This double board wall can furnish quantities of Kriegie stove fuel.

We are jostled and jerked around for a time before the train eases into a rocking gate, clicking along. We are pretty quiet, not moving much. The center plumbing is hardly initiated when the train comes to a smooth straight track, then slows down to a final anchoring stop.

We hear boxcar doors opening, coming our way from the front of the train. Guards' and Kriegie voices sound docile. Our turn finds most of us standing and ready. Jene and I slide on our butts off the doorsill and drop down to the rock ballast. It feels good to stretch and take a couple steps. This seems like a planned stop in an ideal place for a definite reason; a pre-planned rest stop?

The train is standing in bright sun in a rural setting. The roadbed is on a fill that varies in height from ten to twenty feet. Slightly rolling open fields are intersected by the raised railroad bed. The fill extends up and down the track for the length of the POW cars. The front end is to our right as we stand, backs to the track. The engine is nosed barely into a shallow cut in a hill with the soup-food and guard cars at the head of the Kriegie cars. It is a nice warm day in a beautiful spot; deep soil, green shrubs growing at the bottom of the fill thirty feet away, down a steep slope. Most Kriegies are taking care of rest needs alongside the track. Some are making their way down the slope for the "drop-your-pants" type needs. All in all, it is the most space Jene and I have owned since being downed.

Heavy bombers are moving over us, several miles up; not the kind of flying Jene and I did. We are more familiar with railroad fills; implanting bombs into them to blow tracks sky high. We knocked out locomotives standing in marshalling yards, raced them to tunnels, tested their boxcars. Worked on them one at a time; sometimes down the line in bursts searching for cargo types that burn or blast.

One bomber friend tells us the bombers' target is probably Stuttgart or maybe Regensburg; depends on where we are and who is speculating. I do not know our location. We ground attack fighters did not customarily target large industrial cities. There are no black puffs of flak this day. No roaring dog fights. The escort is not visible to me.

As time progresses, more and more Kriegies make their way down the slope; some urgently. Space is becoming scarce at the toe of the fill. We hear the bombing and watch for flak and for hit planes. See neither. Jene and I watch the Fortresses for a short time before turning back beside the car. There isn't a cloud. Bright sunlight cheers us. I am still aware of the high altitude show overhead.

Suddenly, the fighters' engines change tone. I, playing it cool, ask, "Do you suppose they've found something up there?"

I am thinking of an ME-163 or ME-262 attack. Jene looks up; sharply replies, "They have, and it's us!"

They are yet fifteen or twenty seconds out of range. As targets go, we must have been an escort element leader's dream come true. Swoop down and get a fat freight train, no flak, no sweat. The two P-51s come on; we are waiting to wave when one's wings light up. "He's shooting!"

We launch bodily down the slope, rolling flat to the bottom. The fighter fires out of range, walking his tracers to the train. His eyes are better than his judgment. He quits firing just as he gets in range and sights the POW painted letters. The second plane pulls off, wings rocking as he skims over us. The Mustang knocks out our locomotive, and according to the griff, he also wings one Kriegie in the shoulder. We are delayed several hours. The incident would flunk most any test for good strafing technique.

We end up, stretched out flat at the bottom of the fill, heads toward the engine. I have an associated major worry. A couple sniffs signals a probable fouled landing area. How dire? Caution is in order. Hands and fingers OK. Carefully, I raise my head to confirm a safe place to place a hand. Then I repeat for the second hand. I hope to rise straight up, free and unscathed. How about full front and rear? So far, so good. Jene is gingerly checking, too, both still prone. The sense of smell gives us plenty reason to proceed slowly. My eyes confirm contaminant a foot in front of my nose.

Gauging every move, I work in sequence; legs, feet, tuck up; examine each part and placement of same. Eventually, I arrive at standing position. Jene is just finishing a similar procedure with an encouraging result. We check each other. We escape contamination, grateful, and jubilant.

The strafing incident has no excuse. It isn't a case of misidentification. It is not identifying. It became an action without knowledge, AWK! I would like to see that gun film.

Soon we are whittling a handful of Kriegie stove sticks and toasting Kriegie bread with a dollop of strawberry jam, and a couple with chips of bully beef sprinkled with flakes of dried onions. Dried onion bits are between cigarettes and chocolate as being at the top of trading stock. Jene acquired the onions at Stalag IV-B and besides brightening our fare; the result of his trades also adds calories with further barter. He keeps the onion bits in a little pouch like a Bull Durum tobacco sack.

They manage to get out a ration of tea water for all cars. Our time spent turns out to be a recuperating day. We have been very busy the past days, fixing kit, conserving food, and getting ready for whatever. Now, we are into the "whatever."

All in all, maybe the P-51 pilot should be given credit for pulling off in the middle of his run. I never saw the wounded Kriegie. The guards ride in cars at the head of the train. Each Kriegie car has an assigned guard; they are with their cars only when the train is stopped.

During the long wait, our guard is sitting on his helmet just inside the door of our car. Many of the Kriegies are out, alongside. We are not permitted to leave our car area. We cannot walk up to the engine when we are stalled. I am in the car using the guard's bayonet, with permission, splintering off pieces of wood for our Kriegie burner. His rifle is standing in the corner. The Kriegie on lookout signals. The Jerry officers are inspecting the guards. We can hear a German whistle as each car is approached. Our guard isn't hearing anything. He is blissfully sound asleep, roosting comfortably on his helmet. He is jovial, OK. We don't want to see him get in trouble. A Kriegie close by him shakes his shoulder. Hans says, *"Ja, Ja."* The Kriegie insists, *"Appell, Appell!"* Hans opens his eyes. He stands up, sleepily saying, *"Danke,"* straightens his coat and starts down from the open door.

Someone says, *"Eins moment,"* and hands him his rifle. I *"eins moment"* him and put his bayonet into his belt scabbard. Hans *"danked"* again. He jumps down on the track bed where another Kriegie sets his helmet on his head. The roll call is two cars down, getting close.

After the flurry of inspecting the guard is over and the inspecting entourage returns back by our car, Hans comes over to a small gathering of us, saying in Deutsch, and gestures *"Krieg fertig, eins, zwei, drei Zigaretten*-when the war is over, I'll sell you my rifle, bayonet, and buckle for one cigarette each, and I will go with you."

He ends his proposition with a big quizzical grin. My impression seems he is making an offer and request he would like to have come to be, but knows it can't happen.

When we get going, we are not let out for twenty-four hours. In the meantime, we are not strafed though we are at times hearing Ninth Air Force Thunderbolts close above.

Next night, we park in a marshalling yard on the country edge of a city. The guards are giving us lots of slack. After dark, Kriegies begin exploring, sort of like honey bee workers following trails that show the success of the first, and now is returning to the hive. Jene and I have experienced enough tough situations that prove the guards can give protection as well as keep you corralled.

We have been dry but not unbearably thirsty. The engine gives us hot water and ersatz coffee, but water and a container is desirable. We follow the back trail of Kriegies, go across several tracks in the railroad yard. One boxcar has a door partially pried open. A previous investigator has pulled an end from what appears to be rolled woven blanket material. We manage to get a short section out, but then we are stymied. Like a raccoon hunts, we move on, hoping opportunity will give us a higher priority find. We had done OK with one blanket, and in the spring warmth, we'd not have to roll up in it. We still slept head to feet.

We cross the creek on a foot bridge. We can see figures of Kriegies. Three or four are talking with a welcoming lady. Others seem to already have the go ahead. We get a drink from her pitcher pump. We ask for a container to carry water. Jene closes the deal. The best we can do is a one liter bottle. She invites us to help ourselves to her garden, saying about all it has left are some leeks. We trail back to our particular rolling home. Hans figuratively checks us in.

Next morning, Hans is asking people if they have seen his cup. It is a half-liter aluminum, not unlike the GI cups used with our mess kits in Normandy and on bivouac in the States. Hans is puzzled until he realizes he isn't going to get results. His cup is gone. Hans loses a little of his faith in humanity. I do too, when a couple hours later, we are about to unload at Stalag VII-A at Moosberg, one of the Kriegies shows Hans's cup to several of us; kind of a shallow stunt.

Chapter 18:
White Bread

Guards stand outside each car as we form up to march into the camp. Expectations are not as good as ours for Nurnberg. However, we are impressed favorably. It depends on where and when you were captured and how long you have been a POW, whether you find Moosburg adequate.

Shortly before leaving Nurnberg, I examine my right thigh. It is no longer bandaged, but is still filling in. I lightly rub my hand across towards my knee; on impulse, I span my thumb and finger around just above the knee joint. I chuckle, seeing the tip of my forefinger and my thumb. I thought, wow! I look as thin as the starved Ethiopians pictured in the National Geographic. We are thin, and for a short time, lethargic and slipping.

I have to credit our senior American and British officers for rigidly negotiating the psychological high cards they held for us in dealing with the enemy. The local reception managers have us assigned and under shelter in short order. Our first building is just a roof with board fences separating compartments. It looks to be a modified livestock shelter, similar to one for a regional livestock show, designed for a minimum shelter. We are here for only a couple of days while we process, including having sick call and food delivery from a kitchen.

Our first meal is unbelievable. More than we can eat. It is a hash made from Australian bully beef and German potatoes, ladled and handed out, including sliced Kriegie bread and margarine. This is the most rib-sticking meal we've had since Christmas in Stalag IV-B on the Elbe.

Jene and I are assigned to an American barracks. It is a little different than the hundreds of light stucco-coated, drab buildings usually pictured at Stalag VII-A. This barracks is located at the far end of the street about a half mile nearly straight back from the Main Gate. A senior American officer and a couple associates are housed in a large semi-private room in the front section. (Col. Gooding?)

The German Major visits him on occasion. The outside walls of the building are dark wood, not stucco. It sits on 4 x 4 posts about two feet above ground, has ample crawl space available via outside openings.

We move in the 9th or 10th of April. The back fence of the camp is a few steps from our building. Guard towers and search lights reinforce the barrier fence in both directions. Guards with dogs patrol outside the fence. The camp is overflowing with Kriegies of all nationalities. We miss the close association and friendship of our Brit mates (Canada, New Zealand, Australia, and Home Land.)

It is here Army Air Force black flyers appear as POWs. They are from the Tuskegee Airmen flying out of Italy. The three men I know live in a combine. They have been down as long as or longer than Jene and I. We judge so because they flew P-47s on missions similar to us. Their group later flew P-51s in escort. They stand out for their "can do" attitude. Hungry is hungry; OK is OK.

Some of our new mates came directly to Moosburg from Sagan. They all marched the first sixty-five miles together. They were divided at the rail yard, half going to Nurnberg and the other half to Moosburg, so many are now reuniting after a two months ordeal of misery and hunger especially for the Nurnberg bunch.

The guards are not doing much better than the POWs. We are still short on rations; maybe getting enough to stop losing weight, yet are short having enough to gain both weight and energy. No doubt times are going to change, and we want to be in the best condition possible to adapt to that change.

Trade seems to be a possible way to gain vittles. More food will get us thinking again. More brain work will get us more food. Jene and I take stock. We have a few items, like louse powder, dried onion bits, British marmalade, and a few packs of cigarettes. Jene carefully rations his smokes. However, cigarettes are currency used as change or "to-boot" in barter.

I still have both wrist watches. One is GI issue. The other I had bought at the Baton Rouge PX as back-up. Jene did the most trading with cigarettes and food. The gin is you can trade a watch for a Red Cross parcel if you find the right Russian. Jack Lomath gave the impression you can buy anything from the Russians if you have enough trading goods. I eventually believe him whole heartedly. The Russian laborers have sources and access that are unbelievable to flyers. You ask a Kriegie, or go to the fence, meaning wire fence between compounds, or some other source. The contact doesn't take long. Somewhat like the town bootlegger by the post office. He will need to take the watch for an hour. I hand it through the fence and walk on.

When I think nearly an hour has passed, I go back to the fence. There is no trip wire. I know I will be recognizable. My face is easily described. Sure enough, a contact approaches. He indicates it will take a little time yet. I can see his accomplice peddling my watch place to place with potential buyers. I will probably make about half as much as he does. I can recognize trading talent. It is an acquired skill I haven't achieved but know it depends on trust and is stimulated by need and conditions. To be a good trade, both parties must win. Every contact is an opening for more gain through a second or third transaction.

Jene and I each have one unique talent. He is an excellent bridge player, and his work in gas stations boosted his people reading talent. I have driven teams of horses and usually can recognize the back end of one. We forage separately.

Trust pays off. We eat better, have fun trading, and see lots of Kriegies--Turbans from India and kilts from Scotland. We know Brits like orange marmalade; Yanks, strawberry jam. This isn't an iron clad surefire cinch. It means you are careful to ease negotiations until you find the right buyer for your commodity.

A both-wins trade depends on a lot of things. A win-win situation seldom occurs in politics though politicians use the catch phrase to enhance chance to cash in. Too often the informed take a poll of the mesmerized uninformed to ensure objectives of the informed. A simple trade attempts to satisfy both parties and is fun.

We now have bunks, a few bed boards, and a blanket for each of us. This is more than we could dream of a month ago. We share equally.

The griff is read in our barracks about the same time daily around 1700 hours. It looks like the war will end soon.

Messerschmitt 262 jets fly low over the camp almost daily during the middle of April. I saw a couple while flying on missions the first of November. They could leave us in a climb or level. We can out-turn and out-dive them. I have a friend who shot one down flying from a zoom-through on the deck over downtown Kaiserslaughtern.

Jene and I, having capitalized on the first watch trade, agree to go with another trade attempt. We have a little apprehension about the second watch because it is prone to stop. We decide not wearing it might have caused it to get cold and stiff so I wear it continually and resume my trade overtures after it runs steady for several days. The trade goes through very well.

We are beginning to gain energy. The feeling is general. The war will end soon in Europe. The questions are, how soon for us? Will we have to defend, or fight?

Our travels around camp are broadening. Cut strands of wire, like the first cigarette light of the day, just happen and are accepted. Somebody, called nobody, gets the credit; however, no one ever verbalized anybody. The wire has been cut between us and another compound. I stay pretty close to our home vicinity, though.

I see a frightening scene one day. A patrolling guard is walking in the compound along the fence a few yards in front of me. He stops abruptly, turns, looks down a division fence between two compounds. Our compound and the two divided by the interior fence are fairly busy with POWs just moving in the morning sun. The guard raises his rifle, aims, and pulls the trigger. The firing pin is audibly struck, but the rifle doesn't fire. Out of the corner of my eye, I see a prisoner step through the fence about forty yards away. A misfire or empty chamber; I don't know which. I am startled and perplexed. I summarize; reports are of troops advancing rapidly. Kriegies are cutting wires; moving openly. Possibly this guard and maybe all guards are carrying guns cocked on empty chambers. He doesn't react to the loud click. He stares a moment, doesn't chamber a cartridge, and walks on.

A day or two later, Jene and I and most of the men in our barracks are inside following *appell.* The Kriegie barracks watch gives the signal for a German visitor. The German Major comes in and knocks on our Colonel's door. Colonel invites him in. I am standing about three tiers of bunks away, maybe twenty feet. The barracks is fairly dark inside. A small table standing outside our CO's door is visible in the dim light.

Quite soon, our Colonel comes out, closing the door behind him. He says, "Men, I have a request from the Major. I do not want you to react

White Bread

in any manner to what I'm about to say. Do not react."

He continues, "The Major has told me they expect the Americans will take over the camp soon, in days. The Major would like to have the compound in fit shape. He would like to borrow our wire cutters in order to fix the holes in the interior fence. I am going into my room. In five minutes, the Major and I will come out and pick up the wire cutters from this table." He pauses, "Oh, yes, they will be returned."

Our Colonel delivered an excellent script. His own. Everything worked. I didn't see the wire cutters being placed on the table, but I saw the Colonel pick them up and hand them to the Major. The speech and the incident are accompanied by a multitude of invisible smiles. I check. The wire cutters are returned.

Liberation day brought some ego baggage and a truly macho response. The gin is the surrender will occur after a token battle. We will do our morning roll call, coffee etc. and prior to 0800, we will clear the streets. We should get inside, lay on the floor, or we can crawl under the building. Under the building sounds pretty good to me. We are told there will be token fire; stay down and out of the street.

Shortly after 0800, there is firing. Fourteenth Armored tanks come over the hill in front of the main gate. We are somewhat apprised as to what is happening. Apparently our higher-ups had negotiated a token battle and camp surrender. However, the SS Commander got wind of it. There is a contingent of SS troops in barracks just outside at the back corner of camp. They put up some resistance. They are the "egos."

The "machos" come in the form of a four ship flight of Thunderbolts flashing red paint on cowls and wing and tail tips. The sound of those Jugs arriving brings me peeking out from under the barracks. Their thirty-two, 50 caliber guns dampen the SS fanatics' enthusiasm. Thunderbolts fly a classic gunnery pattern against the holdouts. I glimpse the Jugs strafing. Thank God, this time it is Thunderbolts doing the ground attack. I strain to see from under the barrack's edge. I confidently eye two P-47s slanting downward. I know bursts of eight 50s are exploding only a part of a second away. The Jugs pull off and join their orbiting mates. They stick around; I think "just in case," but the situation is cool.

In minutes, the Stars and Stripes are raised over the Prison Camp corner where the P-47s cultivated the "holdout" patch. We get their message. Our troops have moved in. The ground is secure. WOW!

German guards form just outside the fence from our barracks. The guard tower is vacant. Their loose formation, maybe fifteen in all, stand quietly, semi-relaxed, semi-tense, some talking through the fence with Kriegies for maybe five minutes.

They form up to low spoken commands. Their non-com moves them out along the fence by a guard tower and turns the corner at a second guard tower for the surrender point a half mile farther at the gate. I watch them as far as the corner. They seem to diminish quickly. They are marching, maybe downhill for certain. Some are old, some young, some would succeed, some would not. The war will be over soon. Our chances are better than theirs now.

Time seemed to be moving at a snail's pace, however a full schedule of happenings moves along. A Sherman tank, 14th Armored, comes down main street and spins around in front of our barracks. The American flag stays up. General Patton comes rolling up in a Jeep with driver. He has made stops in the half mile between the gate and our barracks. Spontaneously, Kriegies crowd around his Jeep. I don't get closer than thirty yards. Though my vision is hazy, I can see him when he stands up in the vehicle. I hear him speak of how we've suffered, and how proud he is; more necessary things are said. I ask another Kriegie, "Does he have one or two guns?"

Before I was downed, I heard General Patton had given one gun away or, at least, wore only one. The Kriegie says, "He is wearing only one."

The General concludes saying, "You will have white bread before night."

I am sure we will. He has a voice like a controlling football quarterback. You can hear his words. They are loaded with both conviction and assumption. The bread is delivered in perfect condition and before dark.

I know about fresh white bread. I rode horseback to and from school. Mrs. Leth, a very fine lady and friend, kept house for Dad and me for several years. She taught me to play pinochle and made great bread. I carried the wood to the wood box for the kitchen cook stove. She needed extra wood two days a week. These were bread-making days. The loaves

White Bread

were ready when I got home from school.

Mrs. Leth would cut a thick first slice held in shape by the golden crusty heel. Home was six miles from school and five hours from lunch. I had a waistline large enough to tide me over, and here is why--She dappled the warm slice with cold slivers of home churned butter and smeared it with a spoonful of fruit jam. Wowee!

The fragrance, taste, texture, and delight were "sooo-satisfying." The compressed fresh slab lasted only a few bites as I headed out to do livestock chores before supper. Boy, did I have something in my memory to compare with the Four-Star loaf. It stacked up well. General Patton made good, just the way I figured he would. This time everything worked. Many Kriegies compared that first white loaf to angel food cake. I compared it to Mrs. Leth's loaf. Third Army kitchens gave us a big boost, pleasant to both our ribs and heads.

Next day, the camp is in transition, under the control of Allied officers. Our senior officers are wearing neat uniforms. Senior officers in the British Forces are assisted by British ex-POW "bat men."

For a few, it is time to take off and head for old units or to Paris. For most, it is a time to start a healing process. We are told we will be moved as soon as possible.

Dachau Concentration Camp is a short distance away. It is liberated the same day. *I am not aware of it then.*

We each take our turn at guard duty. I serve a tour late one afternoon and evening. My post is near a large tent filled with Army Air Force officers. Things are quiet. I think I will venture a little. I step inside the door of the huge white tent. It is just before dusk. I raise my voice, "Anyone here from Oregon?"

I hear at least three answer, "Yes," among the hundred or so.

My response is, "Let's sort this out a little. Anyone from God's Country?" From downslope in the corner, comes the only returning answer, "Yeah. The Dalles."

We are friends for fifty years and counting. Clair Penners is an Eighth Air Force P-38 fighter pilot. He said, "Between the two of us, we, Dave and I, averaged one good parachute jump." His words were, "My chute opened above 20,000 feet, Dave's as the tops of trees went by."

Clair's 55th Fighter Group moved into our 362nd Fighter Group's *base at Wormingford, England when we moved south to Headcorn preparatory for the invasion.*

Chapter 19:
Barracks Bag

I start letters. My last letter from home was written in October 1944. My last POW letter to home was sent in February 1945. Dad's health isn't good, but he is tough. He and Peggy keep contact by phone and mail. Peggy helps on her folk's farm. I am optimistic about everyone. My first prison camp letters were sent before the Ardennes breakthrough. They should have made it home. My British mates seemed quite positive about mail getting through.

Dad is active, attends high school games, community social and business gatherings. He has bred registered Shorthorn cattle since 1930. In his last letter, he wrote saying, he had sold the cattle, both his and ours. He also says he has rented the cropland to neighboring relatives. He continued, "My livestock is now one rooster, two dozen hens, several cats, and Flicka."

Our little Australian Shepherd puppy is eighteen months old now. Peggy wrote, Flicka and Dad have a cordial friendship. They understand each other. She probably does chicken chores with him. I have not had any home news since their letters written, six and a half months ago. (I did not know it would be still another six weeks.)

I can't start a letter. I keep wondering what to say. In my heart I know Peg and Dad are OK. They have done a lot for the war effort. They are proud of me and I am very proud of Peggy and Dad. She could have finished her degree, but chose to stay on her folks' farm; milking cows, picking berries and beans, helping cook, and visiting Dad. They all buy War Bonds and Savings Stamps. Everyone does.

Her last letter received mentioned she had sent my suit (too small for me) to Russian Relief. The suit would wrap twice around me now. (Twenty-nine letters are returned to Peg, marked MIA.)
John Shelburne, Peg's father, was going elk hunting in Northeastern Oregon. Peg said he is going to see Dad on the way. I feel good about

these men meeting. (They would meet again, soon, when Peg carries the MIA message a month later, December 4, 1944.)

This first letter to Peg is also tough because I have not mentioned my condition other than to say my hand is injured, explaining why someone else writes my first letter. I signed it, with Jack's guidance. Little was said about burns, if anything, other than Dr. Monier's few words about simple surgery when I get to the States. I know I am disfigured. It is something I worry about early on in prison camp. Reality tells me to write. Knowing Peggy and Dad, they will handle what comes.

I have a pretty good mustache. With a little shoe polish waxing, it turns up at the ends. I have no eyebrows and no blinking eye lids; however, I can squinch the remnants partly closed. They are open when I sleep. My forehead has lost its frown and squint creases. The new skin and scar tissue gives me a slick mask-like appearance. I formed an early immunity to my face. What else? I cannot change it here now. One pilot in camp early on with similar burns has both upper and lower eyelid grafts. These were done by German doctors using skin from his underarm. They seem to fold and work well but appear as white patches.

The last of January, I wrote cheery POW letters home. I had a lot to be cheery about. I speculated, "...see you in June."

June is now four weeks away. I've written nothing for ninety days, except the short line notes in Peg's Testament. Nearly a week has passed since liberation. We now have V-mail stationery.

I shave and trim the mustache and ignore what is above. Better yet, I have a third bath in nearly six months. My walking is improving. I am stronger.

Here is Peg's letter from her fast recovering guy:
My Darling Wife;
Sweetheart, it won't be long now. The boys are here and things are looking up. I had GI grapefruit juice for breakfast this morning, sure brought back fond memories. Don't know how long before we hit the States but am set on seeing you in June. I have some photo static copies of the Oregonian which sure look good to me.

My health is very good, however, darling, the docs tell me that it will take some plastic surgery on my face before I do much home visiting. The operation is quite simple but takes several weeks. I'm looking forward to being with you while undergoing treatment. Although I don't look much like David at present I sure feel like him though. Sunny sweetheart, I love you so very much. Please don't worry about me, but thank God for the wonderful blessings he has given us. Gee, darling, I am so eager to hold you tight and kiss you. Six months is a long time not to know what you are doing or how you are, Peg darling. I'm sure that you have received some of my letters. There is so much I want to tell you but I can't seem to put it in words. I've so many plans to tell you and I know that you have many for me. I'm sure glad that we are us. The gov't says that we have 21 days quarantine in a rest home with apts. for married officers, then a 30 day leave at home. Sounds wonderful to me. Give my love to all, and all my love to you, Sunny girl. Yours forever, Dave

I wrote Dad and mailed both letters at the same time. Dad's made it to Shutler Flat via Star Route Arlington a few hours before Peggy's reaches Rural Route Dundee.

The next days seem slow. We are fed. Trading stops. I imagine the value of cigarettes plunges locally in camp, but I have no doubt huge deals are made outside on "the market." One squadron mate, shot down six days after I joined the squadron in Normandy, makes his way to Camp Lucky Strike and processing via Paris with his wit, charm, daring, and trade goods. I attend an evening movie in a cavern-like barn associated with a dairy. It is across the road from the main gate.

We acclimate to the joy of freedom. Confidence is building. We don't have to scrounge for food. Less fortunate displaced persons could benefit from our discards.

Several airports in the surrounding area are supply points. C-47s take Kriegies out on the back haul. In the meantime, we are issued brand new barracks bags for our belongings. On the eighth of May, our barracks is assigned to GI 6X6 open trucks. I don't quite understand the need for barracks bags. Planners probably know best. They do. It is very nice to possess something new and yours.

We load up and truck to a small grass airfield at Landshut. C-47s are unloading and loading quite rapidly but not fast enough to get us on our way before night. We have a few food items, enough for supper and breakfast. No problem at all. We are used to roughing it.

We camp near a German horse drawn unit's supply shed. It contains lots of leather tack and small equipment; flat saddles, canteens, heavy leather lead straps, back packs, and drivers' quirts about four feet long with flat pointed leather poppers. Brings memory focus on a tragic target of horse drawn anti-aircraft artillery in full gallop, on a mission last August. I stuff a saddle in my barracks bag but discard it at the end. I do bring two of the quirts, two lead straps, and a Wehrmacht field pack. It is horse hide with hair on. I also bring two German canteens and a bayonet with scabbard, like the one Hans loaned me to cut wood shavings for my Kriegie burner.

The only bargaining I have to do is with my conscience and the bag's capacity. I figure I need a pack. The expediting articles, lead straps and quirts will be presentation souvenirs. Water holding utensils are handy. The saddle meets the barracks bag's capacity qualification, but I don't bring it.

We make camp on the ground. For something to do, we walk down to the river where it curves around the field. An American anti-aircraft gun emplacement is manned near the end of the runway. We exchange a little talk with the GIs; they figure the war is over.

Next morning, we line up in C-47 size units of ex-Kriegies. I think it is twenty-four. We are about the seventh or eighth unit, waiting our turn. We slept well, are warm, peaceful, and going back! In this case, our destination is Reims, France where we will be pre-processed before shipping to Camp Lucky Strike at Le Havre.

Suddenly, a single enemy aircraft fills the sky with US tracers from gunners of several gun emplacements, including the one where we chatted with its crew the night before. The target is a JU-87 Stuka dive bomber of terrorizing fame in Spain, Poland, and Belgium. Slung under its belly is a brown cylinder not as big around as our 500 lb bomb but a couple feet longer. The plane looks formidable even though it is circling to land quite peacefully. It has fixed gear. Down gear normally signals surrender. A white stub from what probably had been a half bed sheet flutters from its radio mast. In addition to the 20 mm there are probably eight or ten 50 caliber truck-mounted machine guns hammering away. The Stuka goes around once, rocks its wings and comes in again. We figure he decided to take his chances on landing rather than get caught for sure by some extra eager Allied fighter pilot. He makes it in down unscathed in spite of a sky full of tracers.

Ex POWs sweep out to the runway disrupting the flow of liberating

Barracks Bag 129

C-47 flights. Ex Kriegies rush across the grass to surround the "new" Kriegies and "our prize." More stay back; plenty help push the craft clear of the landing area. We watch from a rise along the edge of the field just in back of the spot for our loading formation, eager to keep our turn for starting home.

Chapter 20:
Ray-Bans & Hot Dog

Our turn comes. We load; lift off, not seeing much but feeling a qualified wait-and-see good. We land near Reims, about four miles from where I took off on November 17th, 1944. This is May 9th, 1945: Five months, twenty-six days, and a steak sandwich ago. The setup for processing us is impressive. They move people. We find a bed down spot, and take turns scouting the procedure. We figure the processing lines. One stays anchored in line while the other scrounges for system information.

We sleep in shifts by a hedge. The lines proceed on twenty-four hour schedule. We never stop; we are through Reims processing and on the train in time to arrive in Camp Lucky Strike, Le Havre, late on the 10th of May.

The procedure at Lucky Strike is also well organized. We are rapidly processed, given new uniforms, and eventually a partial pay. I write letters home. We can shower and eat and walk around. Buildings are US Military camp style, temporary single story buildings connected by long hallways covering acres of land.

My eyes are still draining but much improved. I feel soft, but haven't gained motivation or energy enough to exercise. We seem to rest a lot. We are not organized with former barracks units, but do gather with small segments of the past. Quarters are mostly improved tents. Time is heavy. I smoke a couple cigarettes. On the train Nurnberg to Mooseburg, someone always had one going. You could get a light from another's cigarette. No one ever admitted having a match.

Here in Lucky Strike I lighted a cigarette with a match. It singed the center out of my mustache. Instead of the one piece, now I train a singed two-piece mustache. I write of the smoking accomplishment in a letter to Peg. I am not proud of it. It was something I didn't want to do. But I did.

I mention my hair is growing out again from the big clipping we all received: My mustache resembles Robert Taylor's (1940s movie star) and is shaping up quickly. My health progresses rapidly. After the letter of 4th of May, one written via V Mail on the 13th shows all around improvement, yet nothing personal needs censoring before family sees it. It could be my thoughts are purposely suppressed, but definitely, I am in a state of healthy warm-up.

V-Mail: Mrs. A D. Childs
From: 1st LT. A D. Childs 0767034
Dundee, Oregon
My Darling,
Everything's going swell, Darling. We are progressing rapidly towards getting home. I'm beginning to feel quite like a human being again. "I love you more than yesterday and less then tomorrow." (good old Robert Burns) I sure am planning on being with you soon. I sure will be happy when I get on the boat, but it will be heaven to hold you in my arms again. Give my love to Margaret and John and the kids. I have a few small trinkets for the family, but you, my Darling little Devil, will have to be content with me. I'll make a few guesses as to when I'll see you. From the looks of things, I will be in Fort Lewis sometime in the first half of June. It will probably depend on the Docs as to when I get my 60 day leave. I am however planning on being with you soon and for quite some time, say always. Forever,
    David

In a letter the 15th, self-described in the introduction:
"*My Darling, at last a real ole' Davey letter is coming for you.*
Darling, I love you very much and am so eager to hold you close and kiss you!" (Peggy censored the next few lines) *Note: June 2003, still like 'em.*

I am concerned about my eyes. The May sun is bright. The ground is bare; the buildings reflect light. I set out to get sunglasses. I start at Supply. They send me to Medical, who sends me to Optometry. All turn me away with no glasses, with no qualms or ingenuity. If I have a broken leg, I can get a cast, crutches, or even a wheel chair. I can get a back brace, even eye glasses; actually, I am looking for an enviable object. Sunglasses are for "cool", not medical. I should have accepted the eye glasses and reduced the openings to slits with adhesive tape. No one, including me, thinks of that solution. I become a little weary of the each self-exonerating party in the runaround. I need be creative. I amble along a hallway, freelancing for ideas, like going through the discard pile looking for a problem's answer. My idea is to protect my eyes. I ask for sunglasses saying, "I just want to get home in the best condition possible."

I suspect I am talking to 'fat cats' who are stonewalling. Someone else thinks so, too. I walk down a long hallway; a nurse, a blond, catches up with me saying, "Lieutenant, I've got sunglasses. I'll go get them."

I say, "Oh, no. I'll keep looking. I just haven't figured out where to go next."

She is firm. Sounds like my sweet little wife when she knows she is right. "I'm going to go get them. You stay right here."

I stay and she brings the glasses. I thank her. I love her reaction to the runaround I'd been getting. I am really impressed by her kindness and obvious practicality. She doesn't berate anyone. Just leaves. I don't really thank her before she is gone. I never even ask for her name.

Many times I stop or delay writing this saga, but the thought of eventually being able to say "thanks" to her for helping me and Peg to have a wonderful life, starts me writing again. I'll call her Ginny for Lieutenant Genuine.

Other "best" nurses are Anne, Louise, and Marie in Kaiserslautern, also, in Military hospitals in Menlo Park, Van Nuys, Pasadena, and Portland Air Base.

It seems like life is passing us by; we are in some sort of a back eddy. Actually, we are doing quite OK. The war is over in Europe on the 9th of May. We and other thousands of ex-POWs and GIs with points sufficient to go home are plenty to clog the system. Ships also will be loading units to go to the Pacific. It really is a great and well planned effort.

We board the Hermitage, May 22nd, for New York via a one day stop at Southampton, then USA. We are stacked five and six high on canvas bunks on deck. Thank goodness we are on deck. The Hermitage

Ray-Bans & Hot Dog

is an Italian luxury liner converted to carrying U.S. troops. *We came over on her sister ship, the U.S. Monticello, also an Italian luxury liner. The going over was cold, rough, wallowing, and slow, with a freighter convoy. We took our turns at duty below decks with army troops. I certainly was sympathetic with them; cramped, stale air, and seasickness. We were lucky, the four of us in a small state room, only two can stand at once. General quarters sounded several times. Only once did we have surging destroyers dropping depth charges. Candy bars were twenty-four for a dollar.*

Other than the army angel, Ginny, giving me her personal sunglasses, I get no treatment. No one checks my face, not even me. I must have just used my eyes to shave and trim my mustache, comb my hair.

When I first get on board I am assigned a bunk. I toss my barracks bag on it and head for sick bay. I want to get something to clean my eyes. I don't know why other than to get the junk out which includes wild eyelashes growing up under the webs in the inside corners. The doctor is sharp and pleasant; fixes me up pronto. When I stand to leave, hand on the latch, he calls, "Lieutenant, I think you are eligible for first class priority flight home."

I answer, "What did you say, Doctor?"

He repeats, "I think you are eligible for number one air priority transport home."

I answer with a question. "Doctor, this ship is going to England today, and tomorrow will sail for home. Right?"

He says, "Yes."

I continue, "I'm on board and will be in New York in less than ten days. Right?"

He grins and says, "Yes."

I ask, "Doctor, what are my chances of getting air passage and making it to New York by air in ten days?"

He says, "Lieutenant, you've got a good point. We'll be here if you need us."

I thank him and walk out happy.

The trip to New York is wonderfully satisfying. I feel real progress for the first time. Like processing at Reims, we are moving twenty-four hours a day. No subs, no destroyers, just us lining for home.

The first meal is a little exciting. They knock themselves out doing good for ex-POWs. We had been brought up to regular rations slowly and have been on eating what you want for several days. We are in a buffet serve yourself chow line with trays sliding along. We are all pretty well in control, taking servings that fit our stomachs. I am following next to a large framed ruddy faced ex-Kriegie who is selecting his American meal carefully. The piece de résistance is a Coney Island hot dog with all the trimmings. I am salivating as is everyone else. My friend proceeding is obviously a connoisseur for ball park food. He grabs a bun; warm, fragrant, and tender; deliberates, then forks on a magnificent juicy wiener. Boy, it looks good. He spoons a neat serving of relish alongside, then, with only slightly less gusto, he paddles on a beautiful thick smear of mustard. It is a virtuoso's creation. He slides his tray on. I begin emulating his art. Out of the corner of my eye, I catch his first bite right there. Before he gets to the next serving station, he stops sliding his tray and with both hands, lifts his beautiful entree. He bites a full quarter chunk off flush. Then comes the "chomp," the squeeze between roof of mouth and cradling tongue. Oh, the tender taste buds are tantalized, ready with expectation. Then comes shock! Pain! Choking! Gasping, with contorted cheeks turning fiery red. His need is water, room, a towel, maybe a chaplain, and for sure, a ship's larder not stocked with "uncut British Mustard." My unknown predecessor paid the price. His suffering is not unappreciated. We are indebted to him and will always be grateful for his test the mustard sacrifice.

Chapter 21:
DFC and Bath in Tub

The Atlantic crossing treats us fine, although better for some than others. Officers receive partial pay of about eighty bucks and enlisted men one-fourth this amount, not a lot. This is close to what aviation cadets and recruits were paid per month when we started. Soon the money is moving into the hands of a skillful or fortunate few. I don't think rank counts in the redistribution as much as luck and talent. I had played a little black jack in cadets on the train between bases, and again heading for Camp Kilmer to ship overseas, but I figure gambling is not my sport, at least not shooting craps or playing poker. I like to know something about the things I venture money on. I don't even have knowledge of the vocabulary, much less the rules and odds. I watch a crap game by listening. The pot grows until the man with the dice lets his successes ride to $3600. Before his next throw, he pulls in his winnings, leaves his start on the table, tosses the dice and loses.

Seven thousand men; If half are officers and half enlisted and we bring around half of the payroll aboard, it easily can be two hundred thousand dollars being rearranged pocket wise while we glide over the waves.

I still don't know anything about dice and very little about poker. Gambling, yes. Farming and feeding cattle. Marketing crops and livestock involves betting your skill and judgment against politics and weather, and fad or fashion.

I did chance a little in a joint venture. I will call him BJ for Black Jack. He had energy. We played on the Luke Field Alphabetical P.T. Championship team. I attempted black jack rarely, but played on occasion. I saw BJ in action at BOQ and on the train north. His patter and quick wit were fun. If I get the deal I know it will pay to split it with this honest, talented, farm kid from Pennsylvania. First time I won the deal, instead of selling it, I ask, "BJ will you deal, and we split?"

He says, "Sure."

We both make a few bucks. It's great. Like selecting ears for corn-on-the cob, it is so good when it is just right.

When I was on combat leave in London, I met one of BJ's squadron mates. He told me a harrowing story about him. It is a fantastic event of chance with extremely long odds. BJ, like many ground attack pilots, became the beneficiary of fortunate reprieve.

BJ's reprieve comes when his Thunderbolt is hit hard, really hard on a low strafing pass. I don't remember the cause other than he was forced to jump. I believe he, too, was on fire. He was so low when his parachute pops from the pack and the shrouds stream for an instant with BJ at one end, and the unfolding nylon at the other end, a high voltage power line intercepts the forward momentum and BJ whips around or through the cables. The parachute cannot blossom; neither can BJ's feet hold on the ground. He ends with toes tapping the ground between two towers.

Now fifty years later, I am looking for survivors of our class at Luke and settle on the eleven of us on the Championship Team. My source is my battered copy of Class 44-B *"HEADS UP!"* Their home towns are listed with photos and names. I locate BJ's brother one snowy, cold night in Pennsylvania. It is early evening in Oregon, also very cold and snowing. He was feeding hay to cattle. He could have been me or a neighbor. We talk livestock.

I ask about BJ. He tells me, "BJ is gone."

He never really settled back from the war. His liver did him in. I am saddened. I had hopes of hearing his snappy voice again. Confirmation for his dangling in the chute story came easy, yet anticlimactic. BJ made it through his tour, but had a tough time coming home. He remains in memory as one of the many friends who are combat gone.

A top attraction going home on the Hermitage is the First Army Band. The music is marvelous. We keep time; some dance and jive. Music is magnificent therapy. How about "Body and Soul," "In the Mood," "High on a Windy Hill," "Don't Sit Under the Apple Tree," "Red Sails in the Sunset," and "Marzi Doats?"

Jene had a combat story involving him challenging a supposed allied ground controller by asking him to sing a little of "Marzi Doats." Jene avoided an enemy flack trap.

The soloist sings lyrics of the era. His voice is somewhere around the style of Sinatra. He especially put forth when belting out the blues. "Won't you come along with me, down the Mississippi..." I liked his clear style, maybe because Peg and I had lived in Baton Rouge three months while I completed fighter training. We had great memories there; our song even then, was "St. Louis Blues."

The voyage is unremarkable. The weather good; the quarters cramped; the objective terrific. Going home!

I bought a 25 caliber Belgian Browning Automatic handgun, just in case I am expected to have a souvenir weapon. It looks like a mini 45 Colt Automatic. The GI tells me he liberated it from a civilian lady.

On June 2nd, we arrive into New York harbor with the band playing and thousands of ex-POWs and troops enjoying bright sunshine. The Statue of Liberty couldn't wave her torch, it is held so high. I feel so good. We get home in time (to Oregon) to see ourselves in a newsreel, sunning on deck, enjoying the band.

We dock, unload, and ferry across the Hudson to Camp Kilmer by dusk. It is misting. We are separated into groups. Jene and I stand in a formation for a few moments. We wear class A uniforms. Most of us are Air Force with wings. We are the only assembled unit. Small clusters of soldiers in twos and threes walk behind us.

One soldier only a few steps away speaks a little loudly saying, "Look at the Air Force. See that hot shot standing there in the rain wearing sun glasses."

We are "at ease." The individual in charge had just asked us to wait a couple minutes for some logical reason. I know we don't have any "hot shots" in our small group. Then it dawns on me, he is talking about me. My first thought is quick, spontaneous, and easy to do. Simply step from formation, catch up to the smart guy, take off my sun glasses in his face. My actual, more thoughtful, response is to do nothing. He apparently thought we were heading for overseas duty just as he is. He hasn't learned much. I hope he will be OK. Probably has a chip on his shoulder. I would have proven his point and gained nothing. Instead, I learn my burns are not readily noticeable behind sun glasses in low light.

I try to call home. Can't get through. We are given prepared, fill in the blanks, postcards to send to our families. We eat at an officers' mess.

There is music and a few couples dance. I talk with an army nurse. She is heading out for Europe. I grew up going to dances where folks dance almost every dance. Folks did not want anyone to sit, not be noticed. I ask, "Will you dance?"

She replies, "OK."

For me, it is a kind of test run. Would my legs work? She didn't say anything about my face. The music ends. I thank her same as at home in the school or Grange Hall. This kind gal boosts my morale, confirms my trust in nurses.

The military moves us right along. They are impressive, especially when focused. Jene went towards St. Louis; I towards Tacoma, Washington.

Jene is a good friend. He went to hospital, remained in the Service, married, and is soon promoted to Captain then Major. He and I never talked about our part in the war or our units. We have our own survival at stake for the short time we know each other. We exchange letters a couple of times but discussed only current happenings. I called in 1981. Jene's son told me, "Jene died." Found him through a memorized Street number still carried in memory.

Now, I recently talked to his son and wife. It's a marvelous story. They live less than an hour and a half from us.

Jene and Jack Thatcher both asked of each other after the war. Jene and I were indebted to those two Brits named Jack.

Across the continent we roll. I sent telegrams from Cumberland, Maryland and St. Paul, Minnesota. I try to call several times. Always, the train is ready to leave before my call gets through. Across the plains, over the Rockies, day and night, and finally through the Cascades, and to Fort Lewis, Washington; still five hours north of home. They process us rapidly. Again the army works effectively. Doctors examine us. We are issued temporary Identity cards.

When my turn comes to see a doctor, he looks me over, and then says "Lieutenant, you will have to go to the hospital for disposition."

He hit me hard with his statement. I had no knowledge about my family's condition; Dad, Peggy, my Mom? All plans are ifs. It is the eighth of June. I had not figured for this rut on the road home. The doctor is someone who seems might listen. I mentally adjust for speed and deflection and give it a shot.

DFC and Bath in Tub

My throttled down response goes like this, "Doctor, I've been much worse. I've been down nearly seven months. I've been liberated for a month and a half and have done OK without a lot of medical care or even drawing medical attention. Doesn't it seem reasonable that I be allowed a week to see my family and be home?"

I bid low. I figure I could at least go home first. He held me back saying, "Let me check something."

I am standing by a desk in a large processing room. I see and hear ex-POWs from cities and towns all over the Northwest, being assigned transportation and progressing on through. I have given it my best shot. The doctor, my advocate, is gone from the room. He shows up about the time the orders come through for the Portland contingent. My name is on the schedule. Sixty days leave at Arlington, Oregon.

It is nearly dark when we board the Greyhound bus. I relax a little, yet am anxious. How is Dad? I feel Peggy is fine, but want to know for sure. We stop in Chehalis. I get to a phone.

The call goes through quickly. Dayton, Oregon, number 4 x 3. Peggy answers. They will pick me up. I reckon I will be in the Portland bus depot about midnight. My internal clock fails. We seem to be stopping often. It is nearer 2 AM when we arrive at the bus depot. Jack and John intercept me off the step down. Peg finds us coming around the bus. Everyone, John, Margaret, Jack, now eighteen, Mary Ann, fifteen, Peggy and I, all are in the 1939 Mercury four-door.

Peg sits on my lap on the passenger side, Mary Ann is next to us in the middle of the back seat. She never looks our way. Jack is angling his six foot eight inch frame behind John, driving. Margaret hasn't experienced such family togetherness and quiet in the back seat since the "Merc" was new. I'd been dreaming of this time. I still have the glow of Peggy, closely held, in my short list of most precious memories. The trip home didn't take long. Peg's folks are ten or twelve years younger than Dad. I feel even better, hearing he is OK.

Peg, at close range, touched the ribbons under my wings; the Air Medal with Oak Leaf Clusters, the Purple Heart, and the ETO Ribbon, as well as a Presidential Unit Citation and Theatre ribbon with Battle Clusters. She asked, "Which one is the Distinguished Flying Cross?"

I sort of stammer, saying, "Not everyone gets that medal."

Peg answered, "But you have it. They sent it to me. I have it."

Pleased I ask, "What is it for?"

She says, "It is for hitting a ground target with bombs and shooting down an enemy aircraft." (In close sequence, just a few seconds between.)

I am a little disappointed because these are things we are trained for and are supposed to do. I am thinking of a different reaction incident that I feel might have saved two lives; my own, and that of Bob Racine. It will be 1981 before I meet Bob again. It is 2000 before I gathered copies of mission reports from responding squadron mates. No one, including Bob, other than myself, knew my part or the event. It had not been reported. Being alive is pretty rewarding.

A second Presidential Unit Citation had also been awarded to our group while I was a POW. More, including the POW Medal, showed up on Military records.

This wonderful long day is going great. I am swiftly being restored. I would get my first tub bath in six months.

We are twenty-two. It's June. WOW!!

Chapter 22:
Home June 1945

Peggy tells her mother, "David can't close his eyes. He sleeps with them wide open."

John and Margaret drive us to Eastern Oregon next morning. Each has visited Dad at the ranch while I was away. Shutler Flat! The Tree, the fifty-five gallon oil-drum mailbox and turn west onto Bottemiller Road and pass the shed. Mom was born in grandfather Weatherford's homestead house here. We turn north at the cross roads. Home is a mile north beyond Dad's reservoir with its shelter trees, grapes and blackberries. His oasis is watered by overflow when the windmill pumps excess water. Hungarian quail and pheasants love this off limits sanctuary. I ran this mile and finished with a sprint often. This is my best finish ever.

Essentially, I have been away since the end of harvest in 1942. Peggy and I had a month and a half here with Dad, before my aviation cadet class was called up.

John parks under the elm trees by the front gate. Peg suggests she and John and Margaret wait. I go around by the yellow roses to the entrance on the corral side. I open the screen door into the glassed in porch. The guns are all there hanging on the inside wall, my ranch coat and angora chaps are to their right; the ammunition is in orange crate shelves behind the door. Dad's coats are on the end wall beyond the glass door which opens into the living room. Dad is standing by the summer-cold Charm Oak heating stove just inside. I'll bet he saw the Merc's dust plume.

His life was dealt a very tough second wallop last November, following the first, our loss of Mom nearly twenty years ago. Dad and I started our life together on the ranch when I was four and a half. It's been my address ever since.

I go into the house. We hug. I give him a smack. Dad's double role, Mom and Dad, makes a kiss a part of his anticipated greeting. He says "Your smack is as good as when you were a little boy."

We both kind of size up the other for change and health. He starts first, says, "Son, you may have changed. I have a case of beer and several cartons of cigarettes."

Scarce items. I grin. I had mentioned cigarettes in my first POW letters, written by Jack L. We used them for trading stock. His meaning could have had something to do with a deal I proposed to him that first year on the ranch. I remember him telling someone what I had said, "Daddy, if you will stop smoking, I won't smoke when I get big."

Dad quit when I was five. I didn't smoke until I was twenty-two. I didn't "get big" until I was forty-three. The case of beer was about half full when Peg and I left for California. Dad and I enjoyed it some, not as much as I enjoyed summer's watermelons from his marvelous deep-well and windmill gardens. Peggy shared his garden in 1944. We look ahead for one next year. Now, will try for Dad's favorite fishing at East Lake.

Dad is just fine, so is Flicka, Peg's and my little dog, now two years old. Flicka and Lawrence communicate using his voice and her wise response. Dad gives Peg a hug. John and Margaret visit a little while. Peg and Margaret had been here in February. Dad was troubled and had asked Peggy to come help him plan. Peggy and Margaret came by Greyhound. Peggy listened while he sorted things out. Folks wanted Dad to join their bunkhouse crew, only do a few chores, and eat in the ranch dining room. There was no room for Flicka. A partner wished to use our home. A week following their visit, the *Kriegsgefangenenlager* postcard came with the news, I was alive and a prisoner of war, I would be coming home. Dad had hung tight. Peg, Dad and Flicka had decided right.

People are kind to us. Ike, my cousin twenty-five years older than I, invites me to ride on their River Ranch range with him. His wife, "Eddie-Ike," taught all of us younger cousins to swim in the Rock Creek swimming hole.

Ike always tackles things straight on. We are each in a wagon's track at a walk, good for conversation. Their pasture overlooks a few miles of the Columbia River's south shore. It's hot and dry. We are checking windmills and watering troughs, and mostly talking while riding easy side by side. I take off my sun glasses to clean the dried–on tear pools. We turn towards each other. His look is long. Not much is said. He doesn't blink. No one does. No one says anything.

Cousin Frank, the stockman and Creek Ranch operator, had been a barber. He gives me a haircut down at their place. Peggy and I had ridden our horses to visit him and Daisy during the winter we were married.

Peg and I drive into Arlington, my high school town. Our school mascot is the Canadian Honker. I had been presented the school athletic cup for the year 1940. My name is engraved on it with those before and after me. I lettered in football and track four years, basketball two, and band four; held the District mile record.

High school sports were important to Arlington parents and also to the business community. We had a fine coach, Vince Barrett. The school board and the businesses were aware of the importance of education but also how important a winning team could be for business. We won the State B School eleven-man football championship my freshman and senior years.

From the west, the highway turns ninety degrees south up tree lined main street climbing about a hundred feet in elevation before leaving the city limits heading across the Boardman desert to Hermiston. Main street is about five blocks long. At the south end, it crosses the railroad track to Condon just before it completes a climbing one hundred-eighty degree rollback turn. Arriving from either direction, Arlington, with summer shading locust trees, is a beautiful welcoming respite from the bare brown near treeless but gorgeous landscape.

Traveling by train, bus or car, everyone stops in Arlington. Winter or summer, it is quite a place. However, during locust blooming time, the heady fragrance seems to affect normal residents in ways similar to the Mardi Gras effect in New Orleans.

The newspaper editor's wife taught math; a minister taught Latin and government. All eight grades and the high school are in one building. The seventh and eighth grade teacher was line coach. Scotty was a big man with a great sense of humor. He also led a German oompah band. I was a fourteen year-old freshman left tackle. Right tackle was five years older and thirty-five pounds heavier than I. We practiced the left side against the right side offense. Our scrimmage was kind of like flak on a first mission. As I lined up, second play over right tackle coming at me, Scotty asked, "Childs, are you afraid of him?"

My answer was, "No." I think, "Just a little scared."

Dr. Wilhelm, physician and owner of the drug store and soda fountain, had lost a son a year older than I in a car accident. I always got along fine with him except once when he was examining athletes. He checked my pulse and said, "Forty-six, very irregular."

I asked, "Does that mean I have an irregular pulse, or does it mean a forty-six pulse is an unusual pulse beat?"

He growled, "I said irregular."

I never knew what he meant. A few years later, I found distance runners usually have slower pulse rates than others.

From seventh grade through most of high school, I seldom missed an issue of "Flying Aces." It cost fifteen cents and had model airplanes, or "G-8 and His Battle Aces," just a dime; magazines of World War I stories; bought them from Doc Wilhelm.

The first time Peg and I walk into his drugstore after getting home, I am wearing my concealing dark glasses. We sit down at the soda fountain. He greets us and reaches in the cabinet for the gallon jug with orange syrup. He remembers my high school taste for soda fountain orange drink. He sets two glasses with ice on the counter, then adds orange syrup, followed by toggling a soda fountain lever filling the glasses with carbonated water. He sets them before us and says, "These are on the house!"

Mr. Steinke, a neighbor also of German descent, speaks with a definite accent. He and Mrs. Steinke have a large family. Two of their daughters were in my grade and high school. Three, maybe four, boys were in the Military. The first time I see Mr. Steinke, he was with several other people at the Oasis cafe. He greets me, seems like wants to talk more. I like him and all his family mainly because there wasn't anything to dislike. He speaks a few words and seems to be under some sort of urgency. He blurts forth, "Py collee, are all dem peeple badt ober dere?" (Are all those people bad over there?) His question seems to weigh heavily on his mind.

My answer is, "No, Sir. I met some fine ones."

I feel good for the question and better for the answer I was able to give.

We all three bloomed. Dad was so kind to his, a little beyond newlyweds…We service our car, visit a few families in our community; plan our program. Peggy and I love both the coast and the mountains. First, to the coast; blackout curtains are still in effect; beaches are unpopulated and secluded, at least to us on our refresher honeymoon.

The war is still going in the Pacific. Being home is fine, but we still have people scattered coast to coast in the United States and ocean to ocean over the world. I wear my uniform except at home on the ranch.

My 1937 Plymouth, "Miss Willoby," Oregon license plate number 233, previously owned by a state legislator, develops a wheeze shortly before we arrive in Bend. I explain at the dealer's garage, we are on our way to go fishing at East Lake. They say, "We can fix it and loan you a car. When you get back, yours will be ready to go."

Their loaner, a rugged mid-thirties model, worked beautifully up the long narrow two-track trail into the Newberry Crater, past Paulina Lake and the obsidian rubble field.

Fishing isn't the same as those big catches fifteen years before. It is great showing Peggy the pumice slide, Fairbanks Rock, and the big Rainbow trout around the fishing-restricted dock. She catches the biggest fish. Dad really enjoys her announcement, "There is something pulling on my line, and I believe it is a fish."

Dad chuckles as he repeats "--and I believe it is a fish."

Rowing is tough but worth it. Peg's "something" is a fifteen inch Rainbow.

This is only the third week of our sixty-day leave. The trip to Bend and East Lake seems to merge my kid and youth happenings with adult pleasantness. Our car is ready. The people at the garage are very kind. Someone says "Thank you, Lieutenant."

The title, Lieutenant, isn't strange; just a reminder--the war is still on.

We build our plans, sitting at the living/dining room table same as always. Peg and I want to see Mom. She is living in the Pendleton State Hospital. Peggy and I drive up together. I see Mom alone. I want Peg to remember my wonderful mother as described by her peers and in my very early memories.

This visit is OK. She looks well. Nothing is exceptional or has changed.

A few years later, maybe my last visit, she is sitting, speaking to unheard voices; seems conversational, responding back and forth, very low and very fast. Desperate for a connection, I touch her shoulder and say, "Mom, you used to ride horses."

She turns. Light comes into her eyes, clearly, with expression. She exclaims, "Ride astride and go like the wind."

Her words, my treasured words, are the only adult communication we ever managed. She taught me to say, "My name is David Childs. I live at number 15 East 84th street." I was two and a half.

Peg and I go to the Willamette Valley, visit uncles and aunts of hers and mine. My attorney Uncle, Mark a 1918 Colonel in the Field Artillery in France, introduces me to a couple of the Judges of the State Supreme Court in their offices. I think maybe he verbally awards me the DSC, it's a couple notches higher than the DFC. He also insists I be debriefed by cousins. Relatives are extravagant. If the Governor had been in, I would probably have been given the Croix de Guerre.

The only payoff is in penny ante poker with Peg's Dad and her brother, Jack, and a couple of Jack's friends. I don't discourage their assumptions that I must be a poker challenge. I am more like challenged. Anyway, the benefits of spin allow me to bank a quart of pennies now and then.

Peg and I drive our car. Dad drives his car with Flicka. She stays in a kennel in The Dalles. Dad will pick her up in a couple weeks on his way back home after we head for California. Dad has a wonderful time visiting old friends and relatives. We all go clam digging at Netarts Bay.

The war ends. Peggy and I are in Portland when the announcement comes on the front page of the Oregonian and in a streaming banner across the building. The streets are immediately crowded with tens of thousands. Jubilation is rampant.

In the same edition on the front page, a son of an old friend of Dad is listed as killed in action. His name is in a small box bottom center of the front page.

Home June 1945

We say good bye to Dad and Peggy's folks. Dad is staying longer in the Willamette valley. He has a wonderful time; attends Peg's Uncle's farm barbecue and goes back to Portland. He is looking forward to fishing for salmon in the Columbia next morning with his brother, Devere. The ligaments in one knee give way as he steps onto the first step at Aunt Delia's home. He is disgusted with himself, limps back to the car to get a second suitcase, steps onto the first step and injures his other knee.

He goes to The Dalles in an ambulance. We are both in hospitals at the same time. When I talk to him, Dad's only complaint is Uncle Devere caught two fish the next day. He says, "I am sure, if I had been in that boat, I would have caught one, too."

A young surgeon, just separated from the military at Letterman Hospital, fixes his knees. Dad is a good patient, seems well liked. He also persists for therapy. He has treated livestock. He knows if you don't exercise you won't get well. Dad promoted an exercise bar for his bed and later bought a walker for himself.

Chapter 23: Getting Acquainted 1945-46

We head out with orders to report to Santa Monica, California for ex-POW Rest and Recuperation at a beach front hotel. We have time to take the scenic route. Our plan is flexible. It includes sightseeing and visits with families of comrades coming home, and one missing.

Notes from Peggy's letters home:
August 17, Friday--Klamath Falls.
Saturday, 18th, Virginia City, Boot Hill, Carson City
Reno—Club Fortune...David said, "Isn't that Trombetta?" It was Captain and Mrs. Trombetta. (44-B Luke Field) She left on the same flight with me at Baton Rouge. He's been home just seven days, and they are spending their second honeymoon at Lake Tahoe, had just driven up to Reno. Dave was glad Trombetta recognized him.
August 20, Monday--San Francisco--tenth floor of Sir Francis Drake--"Suds in Your Eye"--Fishermen's Wharf--Ice Follies--Oakland, visited close friend and fraternity brother Jim's wife, Janet and baby. Jim still overseas. (Jim wrote letter of encouragement to Peggy when I was missing.)
August 23 (Thursday) Modesto--stayed with Mr. and Mrs. Peck. (Chris's sister, Emma).
Friday (August 24)--lunch with Mr. and Mrs. Christian, Chris's parents--Ethel, Chris's girl friend, came over. Chris named his plane "Lady Ethel" (We would name our son Chris.)
Friday (24th)...Fresno--"Miss Willoby's" gas tank sprang leak after a chuckhole--left her in garage overnight. (Chris, Don, Red, and I joined the squadron together in Normandy.)

Getting Acquainted 1945-46

Saturday, 25th--to Lemoore--visit Maxine Clark and baby, Billy. Don (Clark) is still in France and doesn't know yet if he'll come home or go to China. We saw Don's son before he did. (Actually their ship changed course shortly after leaving Gibraltar; instead of left, they veered right for home.)

Sunday 26th--Santa Monica--Dave is back in the army again--Hotel Club Del Mar--fifth floor overlooking terrace and ocean. Private beach--lounges, bars, swimming pool, dance floor, and--Sunday night buffet-- (Lavish description in letter).

Highlights--CBS--Frank Sinatra, (skinny little guy, sang much bigger than he seemed) Lauren Bacall, Humphrey Bogart, Hollywood--Fox studios and Olvera Street--fishing-Dave caught barracuda--Peggy caught mackerel but won a dollar for first catch of day.

September 10--To Birmingham Hospital at Van Nuys to wait for assignment and transfer to a hospital doing plastic surgery--Bettie (Peggy's Pecos partner) visited--she bought an alligator bag at the PX--Her husband, Jack, is flying in the Pacific area; Hawaii, Tokyo, Philippines and carrying everything from Generals to Christmas trees and Hood River apples.

September 12th--Still waiting for orders at Van Nuys--scarce items such as bananas, cigarettes, and candy bars are in unlimited supply here.

Outside of seeing Captain Blaco, (Pecos Weather Instructor) the detour to Birmingham Hospital in Van Nuys was about the cadence of the song: *"Write me a letter, send it by mail, send it in care of, the Birmingham jail."*

We stay in a motel. I check in daily and finally draw the "get out card." They are great people; hospitals are saturated. Shuffling patients to the best place for their needs is a valuable service.

Dibble General Hospital at Menlo Park is a great assignment; fine orthopedic, plastic, and ophthalmology service. As my British friends would have exclaimed, "They get crackin'."

Housing outlook is bad. Peggy follows leads. Everyone, restaurants, stores, firemen, have leads, and everyone helps.

The hospital, though only a couple years old, has treated thousands of patients. Many come from Letterman Hospital. Trains unload directly at the gate. The hospital received its first combat wounded in February

1944. By June 1945, they had handled their first 10,000 cases.

Nurses and doctors are exceptional. Plastic Ward nurses meet the test no matter what comes by. Their training surely begins in childhood. They are bright, neat, and fun to be around.

The Head Nurse has by necessity the blend of talents of a housemother and those of the 377th squadron's Flight Line Chief Master Sergeant Matthew C. Muldowney. Private Earl Johnson wrote a poem about Muldowney, the verses each ending with:

"These ships will fly and fly damn neat,
By moving hands and shifting feet."

From: 362nd FG History "Mogins's Maulers"

Those lines could be about the nurses.

In addition, a unit of the Women's Army Corps (WACS) is especially trained by Dibble's staff and Stanford University. These young women also fit the mold; do significant and beneficial patient care. Dressings and physical therapy are most visible, however, others work throughout the hospital in offices, laboratories, records and X-Ray.

They apparently have a good time and enjoy their work. Reenlisting rate was high. Their chief loss of personnel was by marriage (From Dibble records).

Doctors, nurses, and WACS of the Plastic Section; all seem to have dual qualifications. First, their specialty and first again, the welfare of their patients and each other.

In a short while, Peggy finds an upstairs two room caretaker's apartment. The main house is brick, is in beautifully landscaped grounds, with a tennis court, lily pool, fish pond, green house, charming outdoor kitchen and living area, all walled in with a large gate opening into a paved parking area. The owner is a fine man up in years. His wife recently died. The estate is for sale. He had responded to the community request for people to open their homes to families of patients.

He says, "We went after the hospital for this area, now let's make them feel at home." Mr. T refused any rent.

Our apartment is a two floor brick miniature of the main house. We live in the two rooms upstairs, former quarters for Mr. T's Chinese cook.

Mr. T has a sense of humor. He comes by one afternoon where I am washing our aged car with suds and sponge. I am in T-shirt and GI khaki sports shorts. He guides a man and his wife, prospective buyers, past me

Getting Acquainted 1945-46

on their tour around the yard, by the fish pond, to the front entrance. I am finishing up car scrubbing when they come out the back door and cross by me to their vehicle. I judge they are in their mid-years, out "kicking tires" so to speak. As they approach, I hear Mr. T. say something like, "This soldier and his wife use the apartment."

When he comes abreast, he says, "Folks, I'd like you to meet Colonel Childs, U. S. Air Force."

I do a double take before I stick my foot in. They beam over this fine "U.S. Army Air Force Colonel." Mr. T. and I both grin.

We enjoy Mr. T's hospitality for a couple of months. However, we keep an eye out for something to rent and finally find a one-room apartment with a huge walk-in closet and a share-the-bath arrangement. With our hot plate and the window sill for a refrigerator, Peggy manages just fine.

We exchanged Christmas cards with Mr. T for several years. We are sorry when one of ours came back marked "Return." He is remembered with a smile.

I am really surprised to find a fraternity brother living in Palo Alto. He comes to his door, looks me over and says, "You're dead!"

We visit a little while. Ned is a genuine first rate enthusiast. His heart and his idea factory are bigger than he. Ned was a foods major at OSC and is now a food inspector for the Navy on ships in San Francisco harbor. He is 4-F because his poor eyesight wouldn't let him qualify for military service. He and his wife, Margie, are proud on-the-way, parents-to-be.

Ned visits the hospital, takes a look at me as we walk on the grounds. My left leg, not quite straight, leads. The right knee, bendable but stiff, follows along. My eyes spill tears. The natural little tubes that siphon tears are high and dry; don't work.

Ned looks at me and says, "Dave, what you need is to play golf."

Man, you can guess, he has been addicted only a short time. There isn't one excuse I don't try. His enthusiasm is genuine. I need him and his remedies. He knows it, and things can only get better. We get "cracking" and they do.

Clubs are hard to find in stores. However, balls are a cinch. Stanford

Golf Course is closed on Mondays. Ned has the morning off. He will pick me up; says wear old shoes and shorts.

A deep ditch crosses the fairway. The payoff comes because it holds about eight inches of water and is choked with water cress. On hands and feet, we crab-walk, feeling and picking our way; what a workout, and successful, too. We glean enough balls to get me going.

We begin with nine holes. Boy, I get physical, mental, and attitude therapy on Stanford's golf course and a couple of public golf courses not far away. Green fees are $2.50 and patients get a 50% discount at Stanford. Green fees at the small courses are a dollar or less. My game isn't great, but my interest is. This is on the job learning.

Ned demonstrates, I follow. I have to adjust to gimmick knees. The ball on the tee is some trouble until I use a built-in guidance system. As I lean forward over the ball, whether putting or swinging, simultaneously my orbs release two droplets bracketing the ball. I can say I have not hit a real golf ball with a real club until my third swing. I play week-ends with Ned. Peggy walks around with me sometimes. In a few weeks, the ward produces a twosome or foursome a couple days a week. We play a lot.

Golf opens a whole new field: Finding things in stores or shops. Chris's sister and husband at Modesto invite us to come duck hunting. We need a trolling motor for our next East Lake fishing trip with Dad. We like records and albums and need something to spin them.

Our immediate search is for records and a Siamese kitten, things high on Peggy's wish list. The rest might be easier with this beginning diplomacy. We spotted the Siamese Cattery sign when first driving into town. This is easy. Twenty-five dollars! I don't try to explain paying for a cat to my father. On the ranch you either trade or raise your own. Dad had a pet cat, "Ole Persh," but Flicka was the only critter ever getting to come into the house.

My first find is a No. 2 wood about ten years old. The next finds are two irons with names like Mashie and Niblick, also a hand-forged steel putter. The irons sported wooden handles of early 1920 or before.

We go looking for a record player. We have "Rhapsody in Blue" and "Oklahoma" albums, and several singles; "White Christmas," "St Louis Blues," "I'll Get By," but we need a player. We buy the kitten; the golf

Getting Acquainted 1945-46

clubs are about $15.00, bag and all. We finally find a home-made box of plywood with turn table, rotor arm, and speaker priced at forty bucks. That is steep! The pawn shop operator says, "I know, but it is down to whether you want music or you don't want music." He is smart.

Peg has bought war bonds, worked, loves music. She has banked our monthly allotment part of my salary, and I receive back pay as a POW. We are not going on a spending spree, but figure we have a little catching up to do. Peg's wishes, the Siamese cat named Sinbad and her new grey squirrel coat in summer storage in Portland, are like saying, "You are wonderful. I love you."

Sinbad is our optimistic, even outlandish, addition to two people who are living wherever they can find a room, but he adapts well. Talks his Siamese jargon constantly, climbs the curtains, refuses to sleep anywhere but in our bed. He becomes a great little retriever. He and I have a running game with a garter. I toss and he retrieves. Sometimes he initiates the game. He entertains us. One other thing about Sinbad: He has more known ancestors than we do. Among them, his Certificate of Pedigree indicates his Great-Great-Grandparents, King Si and Queen A.M. had been imported by Paul Whiteman, noted jazz band and orchestra leader, big name of the era.

Things are scarce. Next on our looking for list is a shot gun and hunting coat and cap. I pay under $40.00 for a new Stevens double-barrel and bring back a couple ducks from hunting at the Pecks for a duck dinner. Ned and Margie have a special recipe.

Oh yes, be on the lookout for an outboard motor. The one we find is a new 2 1/2 hp Evenrude. Dad and I are planning fishing with Peg's brother, Jack, at East Lake.

We have a lot: Each other, music, shot gun, golf clubs, Sinbad, and are about to get new upper eyelids.

We have friends from the past; Chris's family and Ned's family. We are making friends with fellow patients. There is Ludwig, single, and Brad and his pretty wife; both good-looking men, one with a brace and one with a prostheses. They will walk. The doctors split a tendon in one of Brad's legs and install one half in his other leg so he can pick up his toes; he raises cattle and horses in Montana. Three years later, we visit, help put up hay.

Barry, B-26 pilot, burned side of face and loss of ear, and his sweet wife and son: They have living quarters similar to Peg's and mine at Mr. T's. I am standing with Barry in the private lane by their apartment when Barry says, "See that man walking towards us? He's my neighbor and landlord. That's Ty Cobb."

I ask, "The Georgia Peach?"

Barry replies, "The one and only."

Barry introduces me. We shake hands. Barry and I go to Bay Meadows one afternoon at Mr. Cobb's invitation and join him in his club's room overlooking the track. Later, Barry and family visit us at our ranch.

There is Tobi from Florida, single, likes a nurse. We all like her. Alas, she has a friend at home. Tobi is handsome; flew P-47s in the Pacific. He has a facial scar injury. I talked to him years later; he is a produce broker—knows The Dalles cherries.

There is army Captain Clark. He is another great person. Peggy visits with his wife (here at hospital). Peg's notes in letter home: *"--He had lost his lower jaw shot off by a sniper and has been on a liquid diet since February. He's had only one operation in that time, and his pulse went down to 10 on the table. Only the quick action of Captain Patton who put a tube in his throat, saved him. He goes to surgery Wednesday, so Mrs. Clark is worried. He lost half of his tongue, so that is what they are fixing this time--Wednesday note: "He was in surgery 4 hours...took skin from his leg to repair his tongue."*

A P-47 pilot from Idaho was burned similar to me, but still has eyelashes and brows. His before and after pictures appear in the Dibble General Hospital history.

Peg's notes: *"I saw a pedicle today--it was on a boy's face. Pedicles are loose tubes of skin which can be detached at one end and flipped and reattached by this process, and moved again to fill some part that is gone. The tube can be opened and laid down flat (my observation).*

Peg's notes: Husband of a friend Peg rides with, *"She told me, her husband had 52 blood transfusions; 36 in the field hospital in the S. Pacific and the others in Brisbane, Australia."*

I see scenes, one of two handsome young men, each with two artificial arms, chiding each other whenever they allow someone to open a door or button a shirt. The degree of disablement is dependent on each individual's make-up or attitude.

Getting Acquainted 1945-46

October 17, we celebrate a four year anniversary, the date we first met, rally dance at Oregon State, 1941; ice cream cones. This time, dinner out; main course, abalone. Probably won't repeat that menu. Repeated peppermint ice cream cones often.

October 24, First operation: Right upper eyelid: Peggy's letter home- -*David is doing fine--feel sorry for him...has to lay perfectly still on his back until the operation heals, about five days...grafts have pressure dressing...keep him on an absolute liquid diet...no strain...causes blood to rush to the forehead...blood clot could form between graft and flesh...slow and tedious...gave him morphine and Novocain shots, but they sliced over his eye before the feeling was completely deadened. He said it was just like slicing peeling on an orange....met Major Macomber, plastic surgeon--a brilliant fellow."*

November 25, Second operation: Left upper eyelid: Doctors have concern at evening rounds because of bleeding under bandage...Peggy's letter home: *Dave's graft is OK, Major Macomber, Captain Lamp, and another doctor came, 7:30 this evening, I left the room, they removed the dressing, soon as it was off, Captain Lamp came out and said, "*<u>It's all right</u>*."*

Major Macomber told me <u>Dave's eyes had been open so long</u> that there was excess moisture around the grafts which made it difficult. He said, "<u>They won't look as good but will be all right.</u>"

He (Major Macomber) is certainly a wonderful man. Everyone just loves him. He is 38, married, and has a 5 months old baby. (<u>He is one of three foremost plastic surgeons in the world</u>.)

November 26: <u>All is well</u>...*Dave off liquid diet Saturday...I've seen his new eyelid several times...looks good, rugged, but he winks slightly.*

November 29: *Doctors have promised Dave a leave for Christmas...Dave's eyes look better every day. Don't expect to see a perfect set of eyelids, but they are shaping up. He's had a little infection with his right ear where they took the graft. In fact, Lieutenant Peters warned him against duck hunting on his leave.*

The surgery was done under some sedation but with local anesthetic so I can respond to the Doctor's direction. The donor site for the upper right lid was the back of the right ear. The left lid was from the back of the left ear. (Had my ears pinned back by a Second Lieutenant.) The procedure takes an hour and a half. As they work I hear the surgeons mention 20 gage shotgun shells are hard to find. As soon as I am out and about, I find a box of 20 gage shells and go by Major Macomber's office to leave them. He is there, says, "How did you know I needed these?"

I answer, "You talk in my sleep."

Surgeries and sick leaves alternate except for my first two. The findings with the right upper lid led to scheduling the left upper lid surgery immediately. Both are a success although a little rugged.

We go home on recuperation and Christmas thirty day leave after the second operation. I find a dozen wooden duck decoys. The folks at Modesto showed me a system of duck hunting really made for Peg's folks' farm. It is bordered by two rivers. A creek with beaver ponds runs through the middle. Jack and I start our addiction to Retriever dogs and to duck and goose dinners, thanks partially because of my lost squadron mate and friend, Chris, and his family. Chris is now listed KIA.

Jack and I include a couple days of goose hunting on Shutler Flat. We tell Dad we need to check the soil moisture. He smiles, "By all means."

No better way to learn soil structure or hunt wheat field Honkers than digging a goose hunting pit in the field where the geese are lighting and feeding.

Dad joins us in The Dalles on the way back to the Willamette Valley for Christmas. He talks goose pits and soil moisture very well. He also enjoys great food, music and their homegrown Christmas tree. LW brings his walker, (first one in The Dalles) with him and gets around the house well. We sing. LW has a nice baritone voice. Little sister has "White Christmas" in her piano repertoire.

The walker is a real bonus for Dad. Getting it was like moving mountains; bought it himself for the hospital. Used it in his downtown room. Wrote me – *"Had toothache yesterday – found street level dentist. Walked in, pointed at my jaw; he pulled tooth; I ask how much? Doctor says two dollars; paid him; never stopped rolling."*

This Christmas is a time to be happy with thankful memories and plans for the future. We seem to have gifts "a plenty."

Third week in January, 1946: The third surgery is both lower eyelids at the same time. The record includes pathology… *"He presents bilateral ectropion of the lower lids, secondary to burns."*

Dr. Macomber had already done the first two operations and this is his third for me. He did the first two as a Major. This time, he is Lieutenant Colonel, Chief of Plastic Surgery section.

Surgeries are preceded by a presentation before the team of plastic surgeons. The discussion of the procedure makes one feel important in about the same participation as a three year-old while his parent and the barber discuss his hair cut. The actual procedure has elements in common; namely, hold still and listen.

The surgeon pencils in lines on my countenance and outlines the sequence. I pay attention, too. When he says the donor site will be in the clavicular area, I couldn't not interject with, "Doctor, isn't that the area around the collar bone?"

He replies, "Yes."

My high school coach gave us all midterm Cs in Physical Education Health class. "Said we could get As only if we learned all the bones in the body." Today, I timidly ask the doctor if that area has been checked. He indicates that is not a problem.

These lids are partially a success; don't quite relieve all the roll down, but are a tremendous improvement.

Barbers are kind in trimming inadvertent hair from my lower lids and constant growing scalp hair from my eyebrows.

Back to the eyelids--one idea suggested on the ward is to check the record--discover if a foreskin donor site is a possibility. I keep cool, say nothing, although several, some even alarming side effects are speculated.

Dad stays in his digs in The Dalles. Flicka takes up duties on a neighbor's ranch, as a favor to us. Flicka's kennel-life has just ended after six months. This was February 1946.

She and Stock Man had a misunderstanding, but Woman and Flicka nick very well. Flicka wanted to earn her keep. Man tells me, "First morning, I take Flicka out with me to check our herd of fifty 'calvey' registered Hereford cows. (They feed hay on a grass pasture on the valley floor.) We walk among my pride and joys."

Months later, Man tells me the sad result, beginning with, "Flicka *walks out by my side."* He says, *"I speak low, say <u>heel</u>."*

I wince at his word, apprehensive, I listen. When I greeted Flicka, she was a little casual at first, then warmed to me. The Woman said, *"Flicka stays with me. Won't let a man between me and her."*

I wondered why. Now, I will know. Man continues, *"She pricked up her ears, moved by me. I said 'heel.' She looks puzzled, stops, looking back. I command her to 'HEEL'. Flicka busts ahead, bunched the cows. I firmly and loudly tell her, 'HEEL'."*

No mistake for either the dog or the Man. Both were right, each from different cultures. To Western stockmen, "heel" can mean drop back to the heel of the handler. Dad's North Carolina training had taught both me and our dogs, "heel" said, "Heel them up" or "Nip their heels." It signaled to the cows, time to move, get going.

Flicka was taught by Dad such commands as, "Flicka, it's time to go get the cows." or "Flicka, go around," or "Flicka, will you come here, please." Peg's favorite is, "Flicka, will you please take your seat?"

I am very thankful Flicka is still here. I think a lot of Man; I like him. Others might have shot her.

Man continues the story saying she left the cows on the top of the ridge and trotted back down the hill to learn "what next." Before I can explain the difference in LW's culture and Man's culture, he tells me the end of the story.

Man was waiting by the barn door holding an eight-tine basket pitchfork backwards. He said, *"I hit her. She got up. I swung agai*n, *hit her hard, rolled her over. She headed to the house.*

Chapter 24:
Progress

Surgery, March 2nd, 1946:
There is a new surgeon for the team, Colonel Iverson. This surgery is to insert eyebrows. Full-thickness hair-bearing skin is taken from the back of my scalp for eyebrow grafts. Again I hear talking in my twilight sleep. However, this is a minor frustration.

The material for both eyebrows is removed side by side. One is stored. After the first graft is completed, material for the second graft is nowhere to be found. They even look among the discarded sponges. I learn a lot about seeking hide. I am mentally cheering for the search team. However, I am only a little disappointed when they cease the search and take another eyebrow graft from the same site. So, actually, I wear the first and third choice eyebrows. Number three has more of an arched quizzical look than number one. The third eyebrow isn't mentioned in the surgery report. It serves me well, however. Its arched questioning form helps me in determining truth. It asks of statements made, are they creditable or maybe over-spun. Brow number two is found at the completion of surgery. Shucks. Years later, Colonel Iverson received a write up in the Saturday Evening Post. He responded to a letter. He tells me he sees former Capt Lamp and will extend my greetings to him.

The brows heal, look fairly good for awhile but, as expected, shed and finish spotty. I am given a thirty day leave to gain some time and regrowth.

Things look to be shaping for us to enroll fall term at OSU. I plan to switch from Mechanical to Agricultural Engineering to fit with my interest in agriculture and the study of water and soils with associated engineering, agronomy and animal husbandry.

Peg and I purchase a modest two bedroom home, south of town in

Corvallis. We have a garage with an extra room for us to fix into an apartment. There is also a chicken house, well and well house, and room for a vegetable garden. A huge old maple and a weeping willow are pleasant. It is not far from the auction sale yard. There is plenty to interest Dad and Flicka.

Dad's interest in OAC and its fine programs was first brought back to Eastern Oregon by Mother's graduate brothers, and later he led 4-H clubs and was a chaperone for 4-H Summer school on campus. His cooperation with extension service and experiment stations were a big part of his farm and community efforts.

I am transferred from Dibble General Hospital at Menlo Park to McCornack General Hospital Pasadena in May and given a thirty day leave May 29. Travel orders include three meals at one dollar per meal and three cents a mile for travel.

We send Sinbad home by train; destination McMinnville, Oregon. He is too preoccupied with a shipment of birds to look our way when they shut the door in the express depot. He gets home OK: Steps out of his sedan chair, rolls in the dirt, and sizes up the Oregon contingent of his people with whom he shared Christmas turkey; John, Margaret, Mary Ann, and Jack. However, LW is in The Dalles.

Sinbad is a pretty cool kitten. Last Christmas, John and LW chuckled when they went to the kitchen for a little snack. The rest of the family had gone to the show. They met Sinbad coming out of the turkey carcass left on the counter to cool.

Pasadena is "Southern California." Dibble had been "The Peninsula." McCornack General Hospital is in the large hotel on the Pasadena end of "Suicide Bridge" just off Colorado Street. We find a rental room in a private home. It necessitates walking through the owner's bedroom to the bathroom.

Hollywood starlets and celebrities visit patients. One evening, Orson Wells does a magic show for a small group of us seated in a semicircle. Peg is thrilled. She had listened to his Sunday night Mercury Theater radio dramas, including the famous "War of the Worlds" broadcast. We feel the presence of a personable genius.

Progress											161

One afternoon, Peg and I drive east by the towns comedian Jack Benny made famous by often reeling off their names, Anaheim, Azusa, and Cucamonga. Not far from Pasadena, we stop at an intersection where a patrolman is directing traffic. Wasn't unusual. Traffic had been rather light, but pedestrian traffic is heavy. We turn left. A couple miles further, we are driving along a bleak, but beautiful, treeless, soilless canyon. The road is two lane black top. On our left are huge boulders from the size of a bale of hay to bigger than a delivery van. The gradient is fairly steep. An ephemeral stream's bed has absolutely clear water barely trickling off the lip of each rock down to the next. Gully washers seem to have scrubbed and carried away all remnants of silt or soil.

We pass a Fish and Wildlife tank truck going up in low gear. In the next mile or two, we see literally hundreds of young people, parents and children, standing along the edge of a dry water way, holding fishing poles, grips resting, and tips pointed skyward, like waiting for a parade with a balloon on a stick.

We park beside a shallow basin pond of three or four surface acres of water with a dam less than five feet high. The tank truck pulls into the lot, backs down below the outlet, hooks up a large drain hose, lays it in the damp channel, and opens the pond outlet gate and releases the fish into the instant running brook.

This doesn't tell us as much about Southern California as it tells us Oregon is a paradise. The word will get out. Will we be ready?

We join Sinbad first of June for thirty-day medical leave. Jack and I and the new Evenrude go to The Dalles to pick up LW for an East Lake go. He does pretty well on crutches and seems to be improving. Camp is great; Dad manages getting in and out of the boat. No complaints. The little motor works fine. Somehow though, we don't have the right touch. The fish mostly win. On the way out, Dad has an idea. He says, "The creek below Paulina Lake is open season. I have a hunch it might be good."

Jack is eager, so am I. We work our way down to a steep shaded mountain stream. Use flies. Big fish keep little fish off the hook. In forty minutes, we are back with a nice string. We later learn the outlet screen of Paulina Lake had failed for a couple days. Good fortune struck again.

We order a new pickup for Dad. He insists I take his Pontiac, a 1941 two-tone green, five passenger Torpedo sedan with white side-walls. Dad bought the Pontiac because it was comfortable and had a big trunk, easy to put his crates of eggs in and out when marketing them. We trade my car and Dad's old pickup to the equipment dealer for the new one's order equity. It is still difficult to purchase vehicles.

Leave is up and I am on my way back to Pasadena. Peg stays home. She gets our house ready in Corvallis and helps out at home for the summer.

Tobi and I play golf at a course near the Rose Bowl. We also enjoy the hospital's par-three pitch and putt course, the hospital sponsored events, concerts in Hollywood Park, Rose Bowl Fourth of July display, midget car races, Ken Murray's Blackouts with Marie Wilson.

Near the end of my stay, four of us, not too badly beat up, go out to dinner. We make reservations. When the Maitre d' sees us, he quickly reshuffles, and places us beyond a wall; we think from other diners' view.

On the other hand, one Sunday, we are invited by a hotel to give our opinions at a beauty pageant. Seven or eight of us are seated in the front row. We are unanimous in selecting a brunette beauty, about five feet four, nice personality. She tells me her husband is in the Air Force.

War memories are fading with the public. Not quite the same with the hospital staff and patients. We are all anxious to be part of the public.

Note in letter to Peg in reply to her letter of August 2, 1946:

…You are right about it being two years ago, August 2nd, 1944 that I flew my first mission. Two years ago as I write tonight, we are really sweating out a German bombing (Normandy foxhole).

PROGRESS NOTES (hospital)

27 May 1946: *…it is possible that this patient will require eyelash grafts at a later date.*

1 July 1946: *…He has been seen in Plastic Conference and it is felt that a month or more should elapse before any further surgery on his eyebrows is done. However a Z-plasty to both inner canthi can be done at this time.*

7 July 1946: *…the bowstringing of the canthus of the left eye was revised.*

19 July 1946: ...*Patient to have further revision of the inner canthus of right eye in the near future.*

22 July 1946: ...*surgery this date where he had revision of both eyebrows and a free supplementary graft in the right post auricular region.*

3 August 1946: ...<u>*It is believed at this time that the patient will require no further plastic surgery*</u>.

Excerpt from letter <u>to Peg</u>, August 9, 1946:

...I received a letter from Dad last night saying that he was in the hospital for a few days; sort of a checkup. I sure wish I had the low down on it. He said not to worry, also that his knees are doing alright. I reckon that it is his same old trouble only this time I hope that he has caught it in time so that he won't have to spend a long hot time in the hospital. He is in room 208 at The Dalles Hospital.

Saturday night, August 10th, three of us check in at the desk. I am handed a message.

"Call Hospital in The Dalles."

I am worried. Dad has had a problem with fluids crowding his chest since I was about ten. His strong heart along with fine people in The Dalles Hospital on the Bluff had always set him going again.

I chuck stuff into my B-4 bag and head for the Airport. I can only get a flight, LA to San Francisco. I would call Peg again from San Francisco. I am lucky, find a flight to Portland on a major Airline, DC3. Problem; won't take my check. I am in dress suntan uniform, wearing wings, bars and ribbons.

The terminal is one big room. Airline desks line three sides. I start desk to desk, counter-clockwise. "No." "Sorry." "Regulations." "Sorry."

I am wearing a small dressing over one eyebrow, and using the best demeanor I can muster.

At the desk of an airline, having the word Western in its name, the attendant is supreme. She greets me with, "Lieutenant, write the check. I'll cash it."

She barely looks at the check, hands me the $25.00. I cut across the room and buy the ticket. Peg meets me in Portland. We go directly to The Dalles.

Dad greets us. He looks better than I expected. Then, he talks of a will. He has not mentioned the subject since I was about ten. We were doing evening chores. I was feeding my 4-H steer. Dad stopped by the open door. He leaned against the corner next to the light switch. Said, "Son, let's talk a little."

He continued, "If something happens to me, I want you to tell people that I want Uncle___ to be executor and to serve without bond." He repeated the request and asked me to say, "Executor, estate, and serve without bond."

Now, I am twenty-three. Dad wants a will. I call my lawyer uncle. He instructs me to see Sam___, a long time The Dalles friend of his. Peg and I visit with Dad. I call Sam. He understands and asks for current information. He says, "It will be ready by three o'clock."

It is noon. We visit through Dad's lunch. His weeks in the hospital with knee surgery a year earlier, favorably acquainted him with staff. They are friends. Later, we walk downtown, eat lunch and see Sam. His office is across from the Granada on Washington. Sam is kind. The will is simple. This is Dad's sixty-fifth birthday. We cross US-30 (2nd street), and walk beyond fourth street to the stairs at the bottom of the bluff and climb to The Dalles Hospital. I talk with the doctor, it looks tough. We have some time and Dad is resting. We bring a birthday card also Dad's will to his room.

He looks good; sits up on the bed, puts his feet over the side, takes the paper; puts on his glasses, adjusts tray table, and signs the instrument. It is witnessed by staff.

We make a couple light remarks. Dad is doing fine but is tired. We go to the hotel, cautiously hopeful we will win like times before. We haven't snoozed other than on the bus. Next morning, we feel more hopeful; we have not been called. Maybe we are winning. Not so.

He is in a coma, breathing heavily. His strong heart is struggling. This time it won't run. After quite a time, Dad's nurse says simply, "He is gone."

We say a prayer. The nurse's amen is audible.

Dad, Peggy and I settle short of our immediate goal; maybe not far. We three made it for a while. Coming home, going fishing, crabbing, clamming, picnicking, Christmas, and last month's East Lake trip, now, soon heading for school. Since the beginning, school for me has been one of Dad's strongest goals. He has it in his sight. It is up to us to not miss.

His years of service with love are so very remarkable. He did register disappointment at not making the salmon fishing day with his brother, Uncle Devere. I carry his unique vocabulary words and his love for Mother and the land. I hope also for his creed – each person is an individual, responsible for his own merit. Since my freshman term in school we have written almost weekly. Kind of a mutual admiration is always said or implied; always with love.

We make arrangements, call family and take the bus to Arlington. Bud G., a family friend, offers us his car. We find Dad's safe deposit box in Condon and the necessary Bank Manager steelhead fishing on the John Day River above McDonald Ferry at the mouth of Scott Canyon. Mr. B., the banker, is helpful, same as Bud. Both understand.

My minister uncle, Mother's brother, does the service. Dad's brother, Devere, and sisters, Marion and Delia, come from Portland. We return with John and Margaret.

Letters, notes, and cards from folks attending the service and friends and relatives across the country are a tribute to a very fine man. Dad received a like volume of letters and cards a year and a half before when I was down, MIA. Both times he was honored by friends, some not seen in thirty-five years and more.

I am able to get a flight back to Los Angeles. Our check is good in The Dalles and Condon.

My seat mate looks at me several times. I am quiet, remembering two periods; a lifetime and a week. I think of aunts and uncles from both sides of my family and of Peg's family. My life is rolling by in memories' reels.

The man sitting beside me speaks, "You have been burned?"

I agree. We have lots of time. I am deep in rewind when he again ventures, "You know, they have learned a lot with the war. Plastic surgery could fix you up just fine now."

My thoughts tumble. I cage my mental gyro so to speak and only reply, "Yes, I'm in a Plastic Ward at present."

I don't say more. In my head I add up the procedures starting with the caustic pencil under my chin and on my forehead. Is that two? Next is cutting the webs in corners of each eye, also at Stalag IV-B. Two more? I think of still draining eyes, also spill-over tears catching on glasses lens, maybe twice an hour, sixteen in a working day. I add nine full-thickness skin grafts for lids and eyebrows and two revisions of grafts and a revision of the canthus of the left eye. Each donor site is perhaps a procedure also. <u>Just days before, on August third, the Progress Report ends</u> with, "<u>It is believed at this time the patient will require no further plastic surgery.</u>"

August third, Colonel Macomber calls me to his office explaining, "We think we will not risk further surgery."

He asks if I wish ten days leave before I start separation. I respond saying I have enough vacation. If it will help get going, I want to stay. He replies he will have Captain Lamp begin separation procedure.

Evidently the left eye Z plasty result doesn't warrant doing the right eye. Besides I am becoming blade-happy from seeing scalpels at very close eyeball range.

I want to get on with living, going to school, having a family, and ranching.

The suggestion by my seat mate is meant to cheer me. I muse. The war is over. I started at OSC in the class of '45. If I'm lucky, I'll make the class of '49.

PROCEEDINGS OF ARMY RETIRING BOARD FOR OFFICERS
McCornack GENERAL HOSPITAL, Pasadena, California 25 Sep 46
QUESTIONS BY RECORDER: From Transcript

Q. State the nature, duration, and cause of any disability which you believe you have.

A. *I was on a normal reconnaissance flight over Germany in a P-47 fighter craft. At the time of injury we are strafing locomotives at low altitude and high speed. The plane caught on fire from ack ack. When I opened the chute it seemed to injure my knees and the burns are from the fire. I landed in Germany and was taken prisoner.*

Following presentations by Medical witnesses and questions by board members. The Board was then closed for deliberation and having maturely considered the case finds that:

Captain A David Childs is permanently incapacitated for active service.

I was given thirty days sick leave plus ten days extension. November 9th, I received orders for sixty days terminal leave effective 13

November, 1946. "Home of Record Arlington, Oregon" to arrive there not later than 12th of January 1947 "He is relieved from Active Duty as an Officer of the United States by reason of physical Disability."

Last Permanent Duty Station
362nd Fighter Group 377th Fighter Squadron
APO 141 NY, NY

By January 12th 1947, effective date of my Army Air Force retirement orders, we are settled in Corvallis and both are enrolled in school. Peggy is an upperclassman, second term junior. I am an underclassman, second term sophomore.

Chapter 25:
Looking Ahead / Living Now

I make Oregon in two days. Peggy and I are getting ready to go to school. It is a busy harvest time on her folks' farm. Dad's land is leased to our neighbor. The mild moist Willamette Valley climate is good for my recovery.

John and I go to Arlington in his '42 Ford ton and a half farm truck. John tells me, "During the war, shipyards pleaded for trucks." Farm trucks were a blessing. He rented out one of his one winter.

We take time for a short successful China Rooster hunt on Rock Creek where I began pheasant hunting with Dad in 1932.

We pick up Flicka and load our comparatively new 1940 model M IHC row crop tractor. We are selling equipment because people know we won't be using it for at least three years. Flicka gets first class seating coming back with us. She soon adapts at home in the valley. Sinbad ignores Flicka and she vice versa.

The 1928 Hotpoint stove, Frigidaire, and Maytag ringer washer fill the bill in Corvallis. The tractor fits in with John's expanding row crop farming.

I learn a lot about diversified agriculture. Peg's Uncle Urie and John are on the discovery edge of sprinkler irrigated vegetable production in the Willamette Valley. I am amazed at the tons of "roasting ears." They call them "corn-on-the-cob" or "sweet corn." There are bushels of tomatoes, crates of strawberries, and sacks of pole beans like I had never imagined. John also maintains a Grade A dairy. The cows and their feed and pasture crops blend in rotation to the advantage of the land.

There also is Feller, a one-eyed Springer Spaniel. He and Flicka both become duck hunting characters. Flicka was spayed. Dad described in a letter her travels by express railway round trip to the veterinarian.

There are no flood or storage dams on the Willamette or its tributaries. Late fall and winter floods rampage. Uncle Urie got the

Wheatland ferry to haul cattle from the third story of his Grand Island barn with ramps between floors. Downstream at the mouth of the Yamhill River, John's hay barn is three quarters of a mile from ordinary high water. It is our boat landing during winter flood runoff. In September, the Willamette is low. Gravel bars impair log rafts. I begin a lifelong quest for water knowledge.

We are in school and come down to the farm weekends. In March, Flicka would show up missing Sunday afternoons and would be sparkling clean when she came in. I set out to find what she is doing. There are large ponds in the bottom fields left from the high water. Flicka is swimming among ducks. I am puzzled until I discover she is herding little flotillas between connected ponds. Her herding instinct seems to readily transfer from cattle to game birds. She doesn't scare or try to catch, just moves them. I enjoy watching before calling her in from our fun.

When we start school in January, I go talk with Grant "Doc" Swan, OSU track coach. He is kind-hearted, tough, sometimes misunderstood. We talk a little while. Four years before, he had said, "Childs, you can run forever, but let's speed you up. Go work out with the sprinters."

Doc was right. I improved my times, likewise knowledge and confidence. My high school didn't have a track. I trained running on trails west above the Columbia bluffs before school and south along the highway after school. I won the mile against the University of Oregon freshman. I tell Doc of winning races in the service where his coaching paid off.

In May, coach's secretary calls me. Doc wants to talk to me. He asks, "Will you be head monitor for the State High School track meet?"

I reply, "Doc, I don't look any different than I did three months ago. The condition of my face and eyes wouldn't be too well received if calling fouls."

Doc says, "No problem. Monitors will report to you. Jim Dixon is referee."

OSC's likeable and good sized line coach didn't know me but I knew him from way back in my 4-H summer school days. He and a player put on a sham, rough and tumble, wrestling match for our "Men's Smokeless Smoker" entertainment. Jim was line coach for the 1941 team and "transplanted" Rose Bowl of 1942. I was on a track 'ride' running cross-country and middle distance. "Ride" for freshman track meant a job; mine was cleaning in the Gym at 6 am. Shoveling and sweeping tracked in mud from that December's football locker room was a primary task. Sixty cents an hour helped. As a runner, I was in the gym daily for four terms, so knew most of the staff conducting the meet now.

I ran the mile in State High School track meet three times. I knew a little about jostling and regulations. I said, "I'll do it."

Grant Swan gave me a big boost. I believe he had this in mind. I appreciated his thoughtful gesture. At the pre-meet briefing, I learned coaches were complaining of no calls, especially in the hurdles. One of our monitors spoke up saying. "It is the 110 yard high hurdles. Sometimes kids run out of their lane at the finish where straight-a-way lines cross over lines on the curve."

I ask OSC trainer and venerable meet starter, Doc Allman, for advice. He suggests he caution all flights of high hurdle runners to run out straight to the finish line. The meet is going well, not even any problems with a sometimes near bloodletting start of the mile. However, in the last qualifying flight of high hurdles, a young man crosses out of his lane in front of a runner who is knocked into a third runner. The lane's monitor reports the tragic foul. Previous flights had run out smoothly. I confirm Doc's pre race warnings and report the disqualification.

Soon I see Jim Dixon approaching with the Medford Tornado's Coach. Jim introduces us, asks for a description of the disqualification of his hurdler. I respond saying the incident happened in a lane over to my right. Coach, now slightly aggressive, says, "You mean you didn't see it?"

I reply, "It happened two lanes over. Let me get that lane's monitor." I knew Gray from the track team in '42; figured he was cool and straight forward. I suspected they knew each other. I was right. Gray came from Eagle Point not far from Medford. Coach was not belligerent but certainly protective. The foul was accidental. The youngster, a sophomore in his first state track meet and in a fast flight, just blew it for himself and at least one other front runner. Three went down. It had the quality of a flagrant foul even though totally unintentional, sad for him and the boys he took out. The disqualified boy's coach addressed coach Jim Dixon, Gray, and me, saying, "The boy's Dad was behind the monitors on the upslope taking movies of the whole race. You will be hearing from us." We didn't have official replays then. We monitors were pretty sure the movie would back us.

A few years later, Gray and I saw each other at some event. He remembered. When I asked about the film, his answer was, "The developed film came back and the father and the coach took it downstairs to a viewing room and as far as I know it's still down there." Gray added, "The boy became an excellent Pacific Coast Conference hurdler."

Coach became renowned for his coaching at University of Oregon, the United States Olympics track and field team, and developing great athletes. He instigated running for pleasure and stimulated champions to become outstanding altruistic corporate successes.

Opportunities and Possibilities:
In 1945, while still at Dibble Hospital, I received a call asking if I would consider working in the new bank being formed in Arlington. We had been without a bank since the bank closure in the 1930s. I thank them, saying I will be finishing school which will take nearly three years.

Shortly before I was shot down, Dad wrote saying he wanted me to operate our wheat ranch. He also said a man in Arlington, Jim, wants me to rent his nearby wheat land. Relatives, too, offer leases on land.

Another cousin said he was sure I had been receiving good pay. He was an under-water diver and welder on an Alaskan salvage vessel before the war. He wondered if I'd like to loan him some money to help finance his ranch operation since I had not been able to spend much while a POW. We sold him our one and a half ton flatbed truck with stock racks, combine harvester, mulching plow and rod weeders. Instead we got his money. New machinery is very scarce. Ours has been in the machine shed unused since 1942. People are clamoring for it.

Our answer is the same to all. We plan to finish school and will want at least enough land to support us when we come back to the ranch. For the time being, the ranch land is leased and the buildings and house are vacant.

I added up the offers of land generously volunteered in 1945. Together, they amounted to a very nice Eastern Oregon dry land wheat ranch with grazing for a small herd of cattle. Folks were serious but the offers were not open-ended and closed during the years I was in hospital and finishing school.

When we all left in summer of 1945, Dad was coming back in a month and picking up Flicka on the way. We tentatively planned with Dad for him to be with us in Corvallis while we finish school. In the meantime he and Flicka would hold down the fort.

Next time back, just before Christmas, Dad was still recovering in The Dalles. Jack and I checked the buildings, drained water pipes, wrapped boxes of canned goods stored in the cellar with quilts and blankets. We secured all four outside doors, locked the entrance door and left the key in the lock. Six guns, rifles and shot guns with their ammunition, chaps and coats were left in place. I figured if anyone wanted in, they could easily gain entrance. We just didn't want anything broken. Essentially, this was the condition of our ranch buildings from December 1946 until March 1949.

About leaving things unguarded: Items were missed when we came back; a sixty foot, heavy duty three wire electric cord was never found. Our anvil was found but unsuccessfully claimed. Grandfather's 44-40 Winchester rifle was discovered twenty years later, hanging above the mantel of a relative's new home. Last time it was fired, I was near fourteen. We hung it on the east-west barbwire fence by the bunk house. I attached a long lanyard-like string. The front hanger wire held the barrel elevated some. I tugged the trigger string. The hammer fell and ker-boom, out to the west across the county road beyond the canyon, a furrow of dust erupted! Uncle Mark wrote in his family history that I possessed the rifle. The fourth item missed was a child-sized, captain's style high chair. It was old and in good condition, but never discovered. Our house, shop, and barns were stocked with normal utility items found in an operating ranch of the era and location. We appreciated the post war neighborhood's trust and code. We are fortunate.

John Shelburne and his elk hunting mates include me in their fall hunt of 1946, Great camp. I miss a spike. He is practically running over me in the center of a small herd of cows and calves coming downhill in a snow storm. The elk hunt is a nice transition to civilian life. We have great discussions. I do not hunt elk again until we are through school.

In 1949, Lee Pettyjohn and I are fledgling forerunners to a diverse elk hunting group which matures and wanes over forty years. Lee is a former marine antiaircraft battery sergeant in the Pacific. He leases Uncle Mark's land. I apprentice myself and truck to their harvest operation that first year. It is the best thing done in my getting-started. The friendship is strong and lasts. They include me in their continuing discussions of farming and marketing. I essentially am taken in as an apprentice who benefits from their successes and experience, with evolving methods. They are very thoughtful in evaluating, developing, and innovating. Lifetime friendships became a premier bonus.

Classes and farming mesh well. Occasional weekends I drive Uncle Mark up to his ranch near our place. He is Mom's lawyer brother. He also is an observant and innovative wheat rancher. He graduated from Oregon Agricultural College, OAC, and Law school at Michigan, before World War I. He was Cadet Colonel ROTC and became a Major in the Field Artillery in France in 1918. Uncle Mark was a keen student of both the law and land.

In 1932, Mark pulled one of his combine harvesters with thirty-two mules, another with a Caterpillar sixty, and the third with Old Tusko, the first Caterpillar Diesel tractor west of the Mississippi. One of my Ag Engineering professors was the OAC engineer monitoring Tusko's record day and night run. I knew both drivers--twelve hour shifts. It was exciting for a kid of nine.

School is quite different after the war. Before, I directed towards getting enough mathematics and science to qualify for West Point or Annapolis. I wanted to fly and get an education. I still do.

Spring term 1948, I take a course in flight ground school and ten hours of flying at the Corvallis airport. The object is to get a private license and learn to fly a small airplane. Gar, my instructor, takes me around the patch. I listen. I really didn't know how I would be on flying. When we get back down, he says, "You've flown before."

I answer, "I hoped it would show."

I am hooked, get my private license; military time counts. I take my brother-in-law for a flight from McMinnville to view the remains of Portland's Vanport flood disaster. It was another twenty–five years before I would fly again.

The professors in Farm Crops, Soils, Horticulture, Bacteriology,

Agricultural Economics, Physics, and Chemistry open new vistas. We started doing soil tests, paying attention to farming techniques. My last term, I grabbed electives, Photography and Greenhouse Practices. Peggy has the green thumb; I take pictures.

Peg got her degree in Home Economics in 1948; I in Agricultural Engineering in 1949, with honors.

The Dean of the School of Agriculture calls me into his office a couple months before graduation. He tells me he has requests for graduates that I might fit. I say thank you, however our ranch needs our attention. We will try farming first. Weeks later, he calls a second time. I thank him saying, "If we don't try farming now, we won't be able to later." We want to give it a good try. It proved to be a quality life. I finish the degree requirements the end of Winter Term 1949.

I built a light flat bed trailer in Shop Practices class. We bought a used two-ton Chevrolet truck with a combination grain and stock racks. Boy, it has a vacuum two-speed transmission. Brother-in-law Jack thinks it is cool. So do I.

Cousin Ed and I load our Corvallis furniture and drive to Shutler Flat. Ed drives our car and tows the trailer. We leave the truck in the machine shed and unload the trailer. I enjoy the time with Ed. He also is an ex-Army pilot going to Oregon State, plays football and baseball at OSC (Ed became a Forest District Supervisor, but for my money, his forte was with oars and his drift boat through tumbling waters, a virtuoso.)

Peggy and I load the trailer once more. We drive up next day. Our neighbors who rent our land insist we stay a few nights with them. We are on our own by the third day. I go to the hardware store in Arlington. They ask me, "Where is your green?"

I reply, "Green? What could be greener than an over-age college grad starting out farming on Saint Patrick's Day?" It is a great start.

We make a planned return trip to the valley for Peg's mare. Her folks give us a Jersey heifer. Jack gives us a fine young Duroc sow. Peg's Uncle Urie gives us Rock, a buckskin stock horse. We buy his saddle. We also bring two black and white barn cats, Tom and Tomasina, and finally, load one chicken crate holding a dozen hens and a rooster. Our trek is possibly less momentous than our ancestors' a century before.

Looking Ahead / Living Now

Peggy and I put the chickens in their pen. Flicka sits down as we close the gate by the chicken feed house. Peg says, "Why don't you try her?"

I ask, "What do you mean?"

We were with Flicka and Dad in the homecoming in June 1945. Peg also spent time with Dad and Flicka in 1944 and two days in February of 1945. Flicka has been away from the ranch since August 1945. It is now March 1949. Peg cues me. Flicka is patient. I start through the gate to the chicken yard saying. "Flicka, go around."

She hasn't heard the command-request since Dad closed the second hen house in fall of 1944.

Her ears prick up. She puzzles, turns right, then left, utters an anxious little whine and lines out, past the smoke house, slips under the board gate by the root cellar, makes her way through stalks of dry weeds where Dad grew his fine gardens and sits down by the gate next to the high-board corral. I play Dad's part and walk through the main chicken yard, go past the board hatching nests and to the gate where Flicka sits waiting. We walk together on out to the other chicken house.

We are home.

We hire spring plowing. Our renters will harvest their crop.

Peg and I concentrate on changing and updating the kitchen and bathroom. This means a new drain and septic system. The septic system is outstanding because I sketch and build it by the book. After all, I have studied Architecture and House Planning, Bacteriology, and have an Extension Bulletin. People drop by while our project develops. Drain field, tile, baffles, size chamber, concrete or steel, slope depth and all are questioned. But every visitor agrees on one thing. I am doing it wrong.

Squadron mascot and arm patch, cougar on the cloud

"Peg o' My Heart" Red Campbell's plane E4Y
Painting by Audine Campbell

Shutler Flat home circa 1933

George and Maude

David and Cousin Ed 1936

Ed and Dave wheat harvest 1941

David

Peggy

David and Peggy

LW (Dad) and Flicka in Shutler Flat wheat field

Cavy (Mom) circa 1908
"Ride astride and go like the wind"
(Post card)

David and smart dog 1944

Photos 181

Wheeled Luftberry

Atchem, UK pre-combat

Back home, Posey and Peggy

Leaving Atchem for Normandy, A-12, Balleroy
Roy Christian, Don Clark, Red Campbell, David Childs

Marty Lucash, rigger, fits Don Clark's parachute harness, A-12, Balleroy

A Few Pilots A-27 Rennes
Racine – Hogberg – Korosy
Hayes – Ort – Sly – Reed – Johnson – Fairbanks – Campbell – Christian – Clees
Lane – Hill – Childs – W. Davis

"All Came Home"
Bruce – Sly – Fairbanks – Campbell
Racine – W. Davis-Kuntz – Childs

Roy Christian, Red Campbell and 88mm gun-Rennes
---Don Clark pictures

Photos 185

Landscape-on way to Paris
September 1944

Chartres Cathedral on the way to Paris

Eiffel Tower in Paris

Dad in garden pea patch

Dad and Peggy, Shutler Flat, summer 1944

Peggy and sister, Mary Ann

David with parachute and squadron patch, Reims, October 1944

Take off on tar paper

Run up on wire mesh

Roy Barker talking to flight in front of tents

World War I battle field

Wiz Wisdom-Heinkel 111 here before us

Chateau Sept-Saulx, Reims

Dining room of Chateau

Chateau grounds

Jack Thatcher visits Shutler Flat

Easter 1955, Chris, David, and Kathleen

Photos   191

Bob Racine and David Childs, 9th AF reunion, Dayton, Ohio, Sept 1991

Luke Field class 44-B anniversary   Red Campbell, Don Clark and
1944-1994   David Childs 1994

page 216 "Mogin's Maulers," 'tree behind cooking pot.'

Liesel Fromnecht 1944

Liesel points to the crash site 1944

Liesel, David and Uwe Benkel at crash site

Champagne at Romesnil Farm(Chateau des Mogin's Maulers) Balleroy 1994
Beatrice Bouvier-Muller, David Childs, Chuck Mann, Silvie and Francoise Muller, M. Romesnil, Michel Muller, Peggy Childs, and Fern Mann

Beatrice and Mme. Romesnil

David and Peggy, Chateau Sept-Saulx, Reims 1984

Fern and Chuck Mann 1994

Ancient home-built model airplane-France-1929 vintage

Flag raising at Isigny. David, Francoise, Chuck, and Michel

German pill box

Sculpture "The Spirit of American Youth Rising from the Waves"
American Cemetery, Colleville-sur-Mer, Normandy

Wings and Tracks Part 2

Chapters 26 to 45

Returns and Remembering

Chapter 26:
Goose Pits and a Different Wheat Field Set-Up

Vainly the fowler's eye
Might mark thy distant flight, to do thee wrong,
As, darkly seen against the crimson sky,
Thy figure floats along.
William Cullen Bryant

This verse, required for my high school literature class, comes to life on October's Saturdays, while I wait for sound and sight of wild geese, either from a spectacular rock bluff ambush, high above the Columbia River, or from goose pits in the wheat fields, miles back on the plateau.

Mornings, convivial bands lift from gravel islands, and bucking the wind, search out canyons or breaks in the basalt bluffs on their way out to feed, gleaning in harvested stubble or grazing newly greening fields, seeded for the next crop.

I am more pragmatic than smug repeating Bryant's verse when flocks pass beyond my gun's sector. Goose hunting is a part of my heritage. I climbed the first rungs of the windmill's ladder, imitating Dad spotting feeding geese to locate and dig goose pits. I climbed higher on the ladder each fall.

I was nine, used grandfather's Winchester twenty gauge double barrel shot gun, given to me by grandmother. It previously was a birthday gift to grandfather from Mom and Dad before I was born. I hunted cottontails before Dad introduced me to Canadian Honkers.

We kneel in the pit. I hear the swish of wing tips. Suspense builds. They glide in with wings barely stroking. At the exact right moment, Dad calls the shot, low and firm, "Shoot Son."

Goose Pits and a Different Wheat Field

Years later, our own youngsters name binoculars "goose goggles." They place a board across the first brace level of the same windmill tower.

My last goose hunt, like the first and those between, added bonus points to life. I had not hunted geese for years, had sort of quit. A neighbor invited me saying, "Geese are landing. Would you like to join us tomorrow?"

I answer, "Got room for Chris?"

"Sure," came back the reply.

I thank him and phone our son, Gilliam County clerk. We would share the thrill of October's incoming wheat-field honkers.

The setup was not exactly the same as in my youth. I had recently flown over similar goose pits. Looking down on them gave me a visual start. Their camouflaged outlines reminded me of another hunt, distant in both miles and memory.

The first band of geese turned to our call. The wind and cloud was just right. When they are really decoying, it's simple. I fired one chamber of my double gun. My goose falls. So does Chris's. He walks out, picks up our two. The fellows in the other pit retrieve another.

I envision the past; Mallard, my yellow Labrador with his tail over-his-back, high head retrieves. Mallard kept the whole scope in view; decoying flights, downed birds, and multiple retrieves. His ready to act stance never missed a thing.

Goose hunting needs a dog; one nudging your shoulder or knee intensifies the anticipation right up to the shoot command. Reflecting on past hunts and Mallard's alertness brings that other encounter to an involuntary analogy.

The year is 1944: I am escorting my crippled wingman, heading for the safety of Allied lines. We are on an armed reconnaissance mission beyond General Patton's Third Army, hitting southeast of Metz.

The youngster of twenty years is concentrating on keeping his P-47 in the air. Moments before, I gasped as his low attacking plane flips on its back. The wing tip somehow misses the ground as the plane snap rolls upright. His radio quits in mid-shriek. His right wing's gun bay cover, the size of a cellar door, is blasted open. It hangs on, hinged at the back.

His canopy is gone. He is braced in his cockpit, straining. We both are willing the plane to fly.

Wounded? How badly? He is in delicate yawed balance.

I call Red leader, asking should I escort my wingman home. I get back, yes, escort Red Four home.

His battered Jug gains tedious altitude by the foot. The situation is too low and too slow. I swivel neck and eyeball, keep up and out, S turning. Emergency control responds with request for a long count. I get a vector to friendly ground. I act with each thought, "Slow up, get in close, look him over, point the heading, get out to S turn with room to see and speed to protect."

Sliding in slightly above and close, I see into the cockpit. His shoulders are rigid, both hands grip the stick. No blood; he nods OK to my hand pointing a slight change in heading.

I clear and start to move away. My quick scan pours on the adrenalin. Dead ahead, less than a thousand feet below, we are in range of every gun of an antiaircraft battery of deadly 88s with acres of supporting 20mm and 40mm guns, surrounded by nests of heavy caliber machine guns all dug in like gigantic goose pits, engineered for this October wheat field.

No slowed plane can fly through the waiting barrage, or get out of its range. We are fat and vulnerable, like honkers barely stroking, committed to a spread of decoys beyond the camouflaged pits. Best hunters wait until birds are scarcely by, then open up, blasting at short range.

These gunners, too, hold fire. Tracking, they have us. No flock shooting here; each plane is duck soup!

Jolted! I act! Shove stick, trigger down, rudder right, bullets hit like rake teeth; get their heads down. Trigger, wrap left! Plow the ground. Trigger, trigger, sweep right! Trigger, trigger, reverse, low and go, hug the deck, full bore, balls out!

In recount:

The slowed Thunderbolt quivers with recoil of eight blazing fifties. Bullets hail into the gun pocked field. I glimpse helmeted men leaving guns, diving for safety into pit wall shelters. The P-47 bristles and charges the way a razorback sow protects piglets, swinging and slashing in fury.

Each segment of each second is to our good. I hit full throttle, dump flaps, hug the deck; amazed, I see no passing tracers or black puffs, feel no strikes. Thankful, I hunch and fly. Seconds give respite. I catch up to my occupied mate. He is unaware of my yard work. I am smarting and smarter.

He edges home, holding it together, gaining skill by the minute; shakes his head "negative" when I dip a wing, pointing down to an Allied emergency airstrip. Ten minutes more, still braced, he brings the 'Jug' to our home strip. He lands her with yaw, 'barn door,' shell holes, and all: Touches her down at a hundred and fifty mph, same speed she flew home. He lost the canopy. He saved the ship.

I parked and walked the short distance over to Bob and his plane. We checked the big hole in his wing and briefly spoke of his skim-the-deck snap roll, our radio foul up, and his trembling fatigued muscles. Just, casual stuff. We were not surprised to see the rest of the outfit landing before we left the flight line heading to debriefing.

Bob and I played it pretty cool. He knew he was very fortunate. I knew we were very fortunate.

A week before, fighter groups in our French based 19th Tactical Air Command had spent a day's missions gunning deadly 88' sites, to reduce flack on heavy bombers. Then, we attacked with diversion, speed, and surprise. I'm sure "they" remembered. Probably is why gunners streaked for shelter at the instant the T-Bolt's wing dipped. But the split second action came from a half lifetime before, said low and firm, "Shoot Son!"

Another band spots our set-up, turns and sets wings. They stroke lightly for headway; I thrill hearing wingtips swish in soft cadence with a throaty squaw-honk or two. I peek, see heavy honkers. A dozen of them; the coveted prize of my youth. I ground my gun. Don't fire on geese again. The others don't know.

My last goose hunt was the best. I'm thankful for all; first, last, and in between.

"Seek'st thou the plashy brink
Of weedy lake, or marge of river wide,
Or where the rocking billows rise and sink
On the chaféd ocean side?"

In my high school days, I didn't dwell on the meaning of this verse. After I became acquainted with "rocking billows" and a lot more, Bryant's next verse, always reassuring, took on new meaning.

There is a Power, whose care
Teaches thy way along that pathless coast-
The desert and illimitable air
Lone wandering, but not lost.

Bob Racine joined the squadron in Maidstone a month earlier than I. We were in each other's circle of friends. We flew in the same flight in the squadron.

We met again at the 362nd reunion in Chicago in 1981. Bob and I mentioned the event. He told me the CO chewed a little because he jettisoned the canopy. I did not know then or for another ten years that no one knew of my Jug's hog going to war caper. Not even Bob.

After I wrote "Goose Pits," I sent him a copy. He called me, said, "I think you saved my life."

I replied, "Mine, too, with help."

Chapter 27:
Flying Again

A very alert stockwoman sees the swinging back door of my pickup van. I accelerate turning onto the highway leaving The Dalles livestock auction yard. She catches me just after I turn onto I-84. It is sundown, heavy traffic. I am headed east.

Her first honk and blinking headlights grab my attention. I guess the problem, hit the brakes, switch on the flashers and anchor on the road's shoulder; jump out running, slam shut the rear door, and drop the dangling latch pin into its proper slot.

Tires squeal, lanes change. One steer rolls over hard, but springs from under a sidewise, skidding, empty flatbed semi. Nothing crashes; nothing stops, no one yells. Big and small rigs slip by, barely missing the steer and each other. Chagrined, I am frozen for the few seconds it takes to play out the very tense scene. The poor critter, running, clears fast off right into the brushy swale. No wrecks, no one hurt. I feel strongly humbled and grateful. I squeeze between traffic and the rig to peek through the small side door behind the driver; see only two steer calves inside, should have been five. Two have gone cross-country, nearer the sale yard.

Darkness closes down recovery doings within a few minutes. I will come back in the morning. The sale yard offers a saddle horse to use for next day. My old horse, Spike, would not be great around city clutter. I accept.

This personal blunder significantly changes the direction of my life.

Early next morning, we arrive at The Dalles sale yard. The borrowed mare is good for my kind of cowboying. We pasture, feed, and finish cattle year around. Most of the time, I have some critter sense. I do some riding and checking but our cowboying is mostly with gates, feed, and persuasion.

Our son, Chris, is free on Saturday. He volunteers to help. He is good with cattle. Daughter Kit, our cowgirl, is away at school. Before I get to the freeway with my borrowed mount, Chris scouts the acres of industrial land bordering the escape site. He finds a man who shows him tracks where one critter spent part of the night standing nestled near a warm oil storage tank in the tie plant.

I track for an hour through railroad yards, ancient farm corrals, under overpasses, and finally, to a willow thicket where I catch a glimpse of Hereford red. I work him through a willow hobo jungle complete with citizens and all. Out to the east, we quietly move closer. He stops, head up, tail at a questioning angle, standing between a bridge abutment and the railroad track under the freeway. I suddenly am aware of a freight train rumbling, coming up grade behind us. My brain wrangles the train thought briefly. Sound says it is picking up speed. The engines are nearing. I can't spare even a glance. My mount and I accept the train. We eye the nervous yearling. He will break the standoff as soon as his in-process message spurs him. Train, horse, and steer are concocting the action. The rumbling iron and steel accelerates steady and loud.

The engineer is part of the production. He seems to pour on the coal right then. He will haze his side. The steer spooks away fast just as the first engine passes. The cars roll faster than the steer. I give the mare her reins and my heels. We close at full clatter. I whirl my rope. On the right, we are held in line by the train; on the left, by a basalt bank, behind by me.

I flub the throw, miss and pull up. The steer and my horse both act like I have finally gotten smart. They stop, nearly winded. There is an open gap on our left. I figure let 'em stand until the train is nearly by. I quickly coil my rope. I am good at coiling missed loops. But this one came up with a straight end. No knot, nothing to build a loop. It looks like battery acid has eaten a place in the rope. Nearly all the cars rumble by while I knot a gross substitute hondo and slide the rope through into a loop. The last car arrives, the steer gathers his second wind and heads north through the gap. The Columbia River is about two arena lengths away.

In charge of politics, Chris has alerted both the City and State police in case the steer becomes a hazard. He is holding back to not spook the critter. The police cars follow; all are coming up the road to the dam and turn in behind and parallel to our chase. The mare is a willing old girl in full pursuit near where the parking lot for the Visitor Center for The Dalles Dam is today. The Seufert Salmon Canning property had been cleared a few years before. Flat open space is dead ahead of the steer. All of this commotion stacked on top of the shaking up he'd gotten on the freeway urged him to blind straight-ahead flight.

There is only one way that steer will stop before going over the bluff and hitting the river thirty or forty feet below. I have to get lucky. I throw my cobbled loop. It shoots out like it is wired to a guidance system. The knot conglomeration hits the back of his head. The loop flops forward over his nose and snugs up clean around his neck. Good old mare sets up; stops solid. She and the rope hold. I grin.

Straight ahead, across the tailings water, looms the huge powerhouse with the Corps of Army Engineers' Insignia. I reach into my shirt pocket, grab my World War II German camera, bought secondhand in London; adjust the focus and shutter speed and snap.

It has been more than thirty years now, and I can't find which book, I stored the picture. In my mind's eye, the photo has gotten better every year. It showed the steer facing me on the rope, the Columbia's bluff, water, dam, and Corps of Engineers' insignia. I'll grin again when I find it.

This is not the end of the story. We load the steer. The officers disperse. I ride my willing facilitator back to the sale yard. We follow the trail of the other two steers with our Honda 60 mini three-wheeler. I track them up the hill to where they crossed a road into a large Bonneville Power Administration substation. A man stands in a doorway, looking out over the expanse of unmolested soggy wet gravel and loudly warns me to not make tracks over 'his' gravel. Chris comes up the road. We load the three-wheeler.

I have ridden for cattle most of my life. Questionably, I have some cow sense. We decide we best find someone with an idea of the local country. Coming off the hill, I look across The Dalles basin, seeing the river, the bridge, and especially the airport.

Airport! That is the way to see the country! Three stops; first the stop sign, next the toll booth, then the curb in front of the entrance to the flight room. My last piloting flight had been viewing the Vanport flood in 1948. The missed latch caper occurred on Valentine's Day 1975.

"Yep," they have a super Cub. "Yep," they can take me on a reconnaissance.

In twenty minutes, we check the lay of the land, the surrounding ranches, creeks, knobs, and canyons. Cattle are everywhere. This is the time of year when home pastures are used for cattle feeding. I could spend futile days looking. Man, can I see the land. I know I will be flying again and soon. A ground attack pilot's dream! ... shoot with a camera. Oh Boy!

In two weeks, I'm saying, "Peg, the kids are grown. We can use an airplane. Farm equipment parts are getting farther and farther away. Your folks are getting older. We need to see them more. Meetings are ignored that affect our interests because we can't spare the travel time."

She isn't opposed to the idea. I start getting brochures, flight manuals, and even owners' manuals for a couple models of aircraft.

A pilot's license is good for life; just pass the physical, get checked out, keep current, and pass proficiency rides. I feel like a kid; like the days of my youth listening to the radio show, Jimmy Allen and Speed Robertson on the Richfield Flying Club program. I'm a "once was" World War II combat fighter pilot with forty-eight missions and about 500 hours total time. After all these years, "I will get to fly again and learn about land from the air down."

The two lost steers never got to the ranch. They were brought to the sale yard in April. I received a check; probably the biggest percentage gain I ever made on a cattle deal. I was winner all the way.

I took a ground school refresher course. A lot had changed with flying since 1948, all for the better. The crops were in and the cattle on pasture. I had demonstration rides in three different planes. One of my Ag pilot friends located a rugged old bird, a vintage Cessna 182, for us. He and another friend flew me the length of the state, south, to pick it up. With him instructing, I flew it to Arlington from Klamath Falls, the hundredth day after the Valentine's Day steer escapade.

Holding the wheel, not a stick, and with my left hand, was totally backwards to every hour of flying time I had experienced. I sweat some, but it was fun. We flew three evenings dual off his strip before he permitted me to solo and fly on home.

When I stopped flying in December 1994, I had added two thousand hours to my flight log. We had flown several of Oregon's first citizens and planners over her watersheds. Peggy flew with me much of the time. Her folks lived near Salem. I became a farming and ranching participant in land and water matters. Once again, I was a student, hope to always be.

I flew our plane almost daily. We took thousands of watershed pictures. I was over the Cascades the day Mount St. Helens blew. We've used photography to study the nature of watersheds. Most of all, we have learned to read the land and understand a great deal of what we are seeing.

We flew about two hundred specific camera flights, shooting linear sequences and many through 360 degrees. Peggy was with me on this flight in the Washington Cascades looking for beaver meadows. We were lucky finding a series of dams spreading water across a semi-isolated valley. I was eager to photograph this stream reach. We checked for any approach hazards. The air was smooth and I hoped to get great pictures. My scouting completed, I concentrated on a sensible fly-by shooting film, followed by a climb up and out. Climbing out took time.

We headed for Shutler flat. The Dalles wasn't far south of our course. Peg looked around seeing Mt. Adams, Mt. Hood, the Columbia and lots of tributaries. I am wondering because I know something is afoot. The little dear leans up close to my ear; she wants me to hear every word. Oh, Oh, Could be tough. "David."

I am alert, apprehensive; did I push it a little on the beaver ponds? Was the pull up steep? This is unusual. Peggy is really great with flying. It does not matter, rough, windy, low on fuel, she is good, adapts cool and well. What is coming? "Don't you believe we need a camera with a telephoto lens? We would not have to use so much time and fuel setting up for, and climbing back up after taking pictures."

My ears quiver under the head phones. The Dalles is only fifteen minutes away. Wow! What a gal. As usual she is way ahead of me even though we are pretty much together.

Our little airport car runs fine, even the bridge toll has lifted. We leave Melo's camera shop with basics plus a 10 pack of 36 shot rolls.

The events; missed tailgate latch pin, impromptu flight, rope settling around the steer's neck, is most unusual and benevolent.

Chapter 28:
Friends and Memories

This day, May 1984, has been sought by Peggy and me for a long time. World War II has, until recently, been put aside. Now, after forty years, we are taking time to evaluate and look back.

Inexperience in travel shows in our excess luggage.

After nearly a thirty year hiatus from flying, ten years ago, N-34787 became both Pegasus and magic carpet to our ranch and our lives. While we are on this trip, the little Cessna 177 Cardinal is going to have a small camera door installed in the left window next to my shoulder. I am addicted to water, passionate about the study of water infiltrating into ground aquifers and preventing induced runoff from watersheds. Our 180 hp aircraft and camera with telescopic lens are effective implements in studying properties of both soil and water and their self-destruction. Now again, my mission is close support and giving aid to the good guys.

Each time we make this hour long flight, we marvel at extravaganza of plateau, rivers, mountains and forest unreeling below our wings, over the Cascade Mountains into Western Oregon's Willamette Valley. For me, the setting sun shows the same magnificent splendor that it has for a half century. Some of my young boy's imagination still lingers. I see cloud islands, lagoons, isthmuses, and promontories of imaginary places, all vivid in the evening sky.

Tomorrow, we travel Portland to San Francisco and over the top to London. This is my first trip back to a place very important in our lives.

Descending over England's soft spring green meadows and hedgerows, we exclaim at the chrome yellow fields of canola in full blossom. This is far different than arriving aboard a troop ship forty years earlier; no mine fields nor Liverpool harbor. This time, just as exciting and far more enjoyable. I hold Peg's hand as we lean close to the 747's window. I know she is seeing the beloved England she has known only in literature. We are excited! This event, returning to World War II combat airstrips in Europe is to be one of the most memorable experiences in our lives.

I listen to pre-landing instruction detail as intently as for flight clearances. These, too, are explicit. "Pick up your bags, go to customs."

The gentleman at the first desk, dressed immaculately in suit and tie, does not look like a baggage clerk, but I have to get by here to get to baggage. I hand our baggage claim. He is kind, returns the claim tickets and tactfully asks for our passports. The first hurdle is passed. We move up the concourse ramp. Animated chauffeurs extend hand held signs over the ramp's railing. Startled, we read "Peggy and David Childs, 362nd Fighter Group."

This is our man. Our limousine, not just a cab but a long black Daimler, delivers us to the Hotel Rubens. For us, this is "in belly deep grass."

Our New York contingent arrived at the hotel only minutes before us. They look bedraggled, too. They had been delayed on the end of the runway at JFK for three hours.

The Rubens is old and beautiful. Its location on Buckingham Palace Road is historic. From our fourth floor room we look across the street down into the cobbled courtyard of the Royal Mews, the Queen's stables.

By standing on the window ledge and stretching up over the outside balustrade I can see the flag pole on Buckingham Palace. If the flag is flying, the Queen is in. This bit of information is volunteered by our London guide, Maureen.

I think of my departed English POW comrade, Jack Thatcher. For years, he wanted us to go to London with him and his wife, Terry. Now we think of your visits, Jack, and the reminiscing in our ranch kitchen. You were so British even though imitating a Yank's accent. We enjoyed your telling of London, North Africa, Italy, and Australia. Today's hotel lunch of roast beef and Yorkshire pudding would have fed us a week, maybe two, in Stalag IV-B.

We start with a get acquainted stroll on Birdcage Walk to the War Rooms, the super command post of the Battle of Britain. It is heavily protected. Cramped quarters brought memories of austere times. World War II communications, lighting, and warning systems all set the stage for deeper remembering. Churchill's portrait, the war posters, green light shades, battle orders, the desks, the bunks bring recall.

Lt. A D Childs (squadron officer of the Day) will awake Red, Yellow and Blue flights if a 12 ship mission. Also green flight if a 16 ship mission (or another time).

Lt. Childs (group officer of the day) I grinned upon receiving the teletype from *19th TAC ORDER OF THE DAY*

Rendezvous…squadrons...etc. followed by IMPORTANT--Gypsy Rose Lee has taken residence in Reno to have her first baby, write her second book, and secure her third divorce.

SIGNED----

We emerge from this nostalgic window of the past into bright sunlight. Last runners are finishing the London Marathon. Flowers bloom in St. James Park. Buildings with dates in 1600s remind us a portion of the heritage of many of the two dozen of us is nearby. Next morning, we awake to the sound of shod hooves on cobble stones. Horses are in the courtyard across the street.

Changing of the Guard, Fleet Street, Trafalgar square, St Paul's, Westminster Abbey, Tower of London, Park Place, Piccadilly: Names remembered from games of English Monopoly played with Jack, Jocko, and Tex in Stalag IV-B.

I'm also remembering combat rest leave of five days and buzz bombs. I stay at the YMCA. The sleeping quarter is a large room with rows of canvas folding army cots. The first night, I slip between smooth fresh sheets, breathe deeply and totally relax. I am well on my way to a great all night snooze when I hear a faint putting. It sounds like a youngster playing in the dirt with a toy car. The sound grows louder and more distinct. I feel something is headed our way. Buzz bomb? I deduct "Yes." Well, no one is leaving; air raid sirens are going. I'm remembering stories of Buzz Bomb strikes and their passing over Maidstone. Bill Fisher has three or four kills painted on his "Shirley Jane." We know they travel around 400 mph, don't fly very high. They plunge to earth when they run out of fuel.

I listen. Even a greenhorn would guess that it was good for the individual to have the 'putting' continue until it gets beyond one's self. I relax slightly when it passes. Then silence. Oh! Oh! Then, Boom! Not far! I listen and sink down in nothing to sink down into. After three or four of these sequences, one quit before reaching us. Wow! I really sink so deep that when the not-close bomb goes off, I catapult up two feet above my cot by the built up compression energy I have transmitted while sinking. This, of course, is green horn reasoning. Kind of like fox hole rotation in Normandy.

I attend plays and shows, "Blythe Spirit" and "While the Sun Shines"; Madame Tussaud's Waxworks; Phyllis Dixey and the Windmill; my ordered red silk-lined tailor-made Eisenhower jacket is picked up by squadron mates on next leave. I also bought a second hand German camera; dined at the Savoy, and made purchases for Peggy; an English riding crop and spurs, a three volume set of a Sir Walter Scott First Edition. These all are carefully wrapped to send home. I especially remember the "young lad" up in his eighties, who prepared the package for shipment. When I returned, the package was ready. He is finishing what looks to be his lunch of salt and pepper on a nice boiled, jacket-on potato. He leads his bicycle through the door, down one step to the alley. His departure is measured and steady.

At day's end, Peggy and I attend vesper services at Westminster Abbey.

The Rubens concierge suggests The Baker and Oven for dinner. Eight of us are seated in an arched alcove, an old oven similarly shaped to the arch of an ancient stone bridge. The ceiling is so low, we can hardly stand. We enjoy warmth and friendship. Peg has steak and kidney pie. Jocko, the Scots RAF Halifax Gunner in the bed next to mine in the Stalag IV-B infirmary, said those words, "Steak and kidney pud," so lovingly, as though he was talking about Thanksgiving turkey with all the trimmings.

Next morning, the Queen's flag is flying. A continental breakfast is served with a gentle wake-up knock. The jam is black currant.

Down the lift with a little time to spare, we cross to the Mews and around the corner, along the brick wall of the Queen's garden. A sparrow evades my camera and flits over the iron spikes topping the wall. We hurry back. We are anticipating our first visit to a former base of the 362nd Fighter Group, Ninth Air Force.

Friends and Memories

We meet Arthur, our driver; a Cockney, born within sound of the Bow Bells. He is waiting to take us to Wormingford. We are getting to know everyone. What a marvelous group; Pilots, crew chiefs, public relation, adjutant, armament, motor pool; wives, a son and daughter, and one grandson.

Chuck, pilot, Luke Field '44, and his wife, Fern, younger than most of us, have planned well. We will be visiting three air fields, and five airstrips the 362nd flew from; two in England, four in France, and two in Germany.

Today, it is Ed Maclean's show, also Nat, Bob, and Mani. They were stationed at Wormingford. My connection is remembering the times as told by the "old pilots."

We drive through London suburbs, exclaiming at names of country pubs. We pass the "Frog in the Nightgown" and on to the "Shoulder of Mutton." There we meet Derek Webb, assistant to the mayor, who escorts us to the Wormingford base.

The field, now farm land, is slightly sloping with a gentle roll at the edges. We drive on the taxi strip around the perimeter. Good looking wheat grows on either side. Not much is said. The louder ones of us are respectfully quiet to those who have been here. We drive around to the end of the cracked runway with its crumbling edges. Arthur stops the bus and silently opens the door. We get out and take a few steps, rather like a small band of grazing Canadian Honkers settle down and spread out after landing in a wheat field.

In the background, a sound from the past breaks in. Again, like wily old geese, we focus together on a huge four engine propeller driven plane coming out of the mist, low and directly our way. The happening is real isn't it?

The timing of entry and exit is precise. I snap three pictures. The craft shows well, not ghostly, perhaps heavenly. The aircraft does a one-hundred and eighty degree turn right over us and disappears back into the haze. The event is as unusual as going to Lochness and seeing Nessie.

I walk on the old runway and out into the green wheat. I see the soil planted for centuries. It is sprinkled with small white stones. Our land on Shutler Flat became a century farm only a few years ago. Clair Penners, a fellow POW, a P-38 pilot from the 55th Gp and The Dalles, flew from here after the 362nd moved to Headcorn.

We drive the perimeter road around growing wheat; we pass the farm buildings, stop and chat; walk by the blossoming apple trees, and look into the old bomb shelter, still standing.

We lunch at the Siege House in Colchester. Its three hundred fifty year old bullet holes are from a civil war predating Gettysburg by two hundred years. Derek Webb is kind and helpful. I first experienced that British ease and charm on my third mission, close behind the battle lines in Normandy.

We chased enemy aircraft beyond Paris. I am flying Captain Frank Pepper's plane--remember my mentor, Pep? "Keep the RPM up, and she won't get a flat shaft." I am tail end Charlie – on outside of turns, use more throttle and also follows instructions, "Keep the RPM at 2450."

Lt. Childs lands on a partly completed British strip as a steam roller takes evasive action. The Thunderbolt's engine quits, out of fuel on roll out.

I am treated to lunch with the Chaplain, while the good people give me 100 gallons of 100 octane from jerry cans at 5 gallons per can. First time I have a Chaplain to cheer me up. I feel much better when on takeoff, the other three ships of my flight come up off an adjoining strip, freshly fueled too.

Colchester Castle is very old. The well and displayed charred boards that once lined its inside were in use fifteen hundred years before. The Castle as it stands with its dry mote surrounding, brings thoughts of battle, siege, resistance, progress, and history. We do make progress, don't we, between our lapses of civility?

A beautiful green plumed domestic Mallard drake waddles his strut, over the trimmed grounds. He could be a distant cousin to the ducks back home, nesting on our ponds. Perhaps, we too, have much in common with our ancestral land.

Chapter 29:
London Maidstone

I checked the flag pole. The Queen is home. Peggy and I are on the curb before the coach arrives. Next door, in a little shop with camera and tourist needs, I strike up a conversation with the proprietor; also purchase a light telescoping tripod. I find him to be a World War Two Spitfire pilot. We team up for a practice remote portrait of Chuck, proprietor and me.

Arthur is in top form saying, "I'll give 'em a blahst on me 'ootie."

We drive out of London. Coventry, a name on a road sign, stirs memories of headlines in my youth forty-four years ago. The Battle of Britain started that summer. It pitted fighter planes, Messerschmitt against Spitfires, JU-88 bombers against citizens. It wasn't called World War II yet. I was in high school. "Flying Aces" a 15 cent pulp magazine featuring World War I combat between Fokkers and Spads over the front lines was my recreational reading.

Traffic slows to a creep. Time enough to remember. German Bombers obliterated the city of Coventry in November 1940. I look out into the thinning congestion of London's outskirts. Arthur drives up a few lengths at a time. November 1940; I am a high school senior on our small school state championship eleven man football team when Coventry is destroyed.

We near the cause of our slow up; a lorry is slant parked by a mangled bicycle. Close by, a brown tarp covers the non-survivor. Forty-three thousand civilians were killed and fifty-one thousand were dreadfully injured in the Battle of Britain and the London Blitz, June 1940 to May 1941. I ran the mile in State High School track meets in May both years. I wasn't thinking much about the war in Europe.

Beyond the suburbs, we drive by fields of canola and grain. Flocks of sheep graze.

Headcorn with its old cemetery, church walk, and its friendly people in shops and on the street is our first chance to really mingle in a British village. Lunch at The Beam; Peggy reveled in the attention from the three immaculate waiters. This was not your ordinary Oregon early afternoon luncheon. Seemed like much more than just three courses. We are still on our shakedown association and everything is in the green for go. It is a cinch, however, we are spending too much time for lunch dining. We soon learn to adapt. Catching impromptu lunches becomes great fun.

Later, we will reorganize, requesting a French bread ham and cheese long boy sandwich. I won't try to translate it. But I'm sure it will become a phrase as much used as "Box la Rue," an attempt to adapt to French language the American slang – "hit the road." I witnessed the beginning of this phrase from conception to infancy, in the middle of the dirt street at the Rennes Air Base, A-27, between Alan Campbell, Ed McClean, and a Frenchman. The discovery was probably not unique with the 377th squadron, but we thought so; certainly was animated discussion including a fist pounding in the road dust and a quizzical raised eyebrow under the Frenchman's beret. *"Whas ziss say? Ziss Box La Rue?"*

The Lashenden Air Warfare Museum is an unplanned for pleasure; including its exotic chickens in the farm yard and the displays of World War II memorabilia. The buzz bomb with its black swastika gives me a chill. Not like an ME-109 in the air or the clang and clatter of tracks on a Tiger tank, but more like the sudden silence in London when the "put put" of a buzz bomb's engine stops, until the explosion.

By chance, we are at the right place, furnished by caring and remembering people. The proprietor of The Beam had suggested we visit the museum while he rounds up our luncheon materials and his crew. I am drawn by an R-2800 P-47's engine, its prop blades bent, also a "Jerry can" (five gallon container) with our 362nd name clearly stenciled. We pose for pictures; run our fingers over the bent prop, and look at the big radial engine, its block still dripping oil after all these years.

Weeks Farm, site of Airfield 412, is delightful with its old half timbered house, rambling gardens, mossy brick walks and statues; the

London Maidstone

Colonel's digs in 1944. This is MacLean land also, but it is Paul's as well. Back then, Paul is a tall, young, towhead with a southern accent; he was the Group's Public Relations officer; likeable and sincere with lots of depth for his years. I first remember him at Group briefings at A-12. Today, Paul and Mac look like pups lolling on the lawn.

I have never seen a retriever hit a scent any better than Big Mac when he starts looking for the pond near where the runway had been. Ed MacLean is the all around biggest fighter pilot I remember seeing. Mac never breaks stride, through the board gate, across the pasture, me in hot pursuit. Mac saying, "I'll know I'm there when I see the pond."

I'm looking at fertilizer spreader tracks on this good-looking pasture. I also see a break in the tree line where the runway once was. Then old Mac comes to point, "There it is, the pond!"

Boy! The look on his face. I beam too. I honored his point. I wasn't stationed in England but joined the group at A-12 in Normandy. We are pointing like a pair of bird dogs out by a brush swale in the middle of a sheep pasture south of London. Again, I could faintly hear the engines. Nostalgia and imagination combine for a mix of stirring pride and warm gratitude for being here in the present. We walk back across the pasture and through the gate.

The bus is loaded when we get back, but people are patient. Seems like a special kindness and understanding. Later, I would understand even better when we get to my first airstrip.

We gather at the RCAF commemorative marker which has since been enhanced by the 362nd. Paul's second memorial service is simple and direct. I'm sure everyone feels as I do that Paul's accepting the memorial assignment and the manner in which he carries it out is a vital part of the success in our venture. It is mostly Protestant Christian, but the meaning could be understood in anyone's heart and heritage.

On to Egerton and the George Inn, seeing the old stone built inn and meeting its proprietor, a British contemporary, Mr. David Drew. At the time we arrive, the inn is closed. A huge sounding dog is barking from within. Mr. Drew invites us in, gives us the run of the pub. It is fun looking at displays of memorabilia and most of all, talking with Mr. Drew.

The classic fighter pilot's hand positions, the animated discussions between P-47 and Spitfire people, the poring over maps is almost like a debriefing. (I think of Lt. Smiley or Capt. Alan Campbell as we unloaded with our stories at intelligence debriefings after missions.) From the dart board on the wall to the model of a Spitfire hanging over the heads of those doing daring dialogue, it all ends too soon. Throughout our trip, we are not dispirited when leaving one place because of anticipation for the next. Going to Hythe is no exception. We see stony soil, a windmill, and heavy power transmission lines with flocks of grazing sheep underneath. I think of steam trains and passengers on their way to the Channel; before that, horse drawn coaches. Centuries have gone by; the country and land is still used. The soil is fertilized; sprayer tracks show in the grain fields. The crops look good.

Peggy's and my family names of Childs, Carpenter, Sperry, Weatherford, Alderman, Shelburne, maybe Odell, seem to denote our people were here.

Hythe, a town on the Channel, is appealing. Inquiry directed us to the Imperial Hotel. A man says, "You will know it when you arrive."

He is right. The old hotel is magnificent. Its appointments outdo the Heathman of Portland or the Davenport of Spokane, high points in my prewar recollection. Peggy, it's also nice to think of those other good times. We've lived in apartments smaller than our bathroom in the Imperial. Housing was tough near military bases during training and while in hospitals after the war.

Fern, Peggy, Chuck, and I walk along the seawall and down to touch the water of the English Channel. The beach is made up of small sea smoothed pebbles. I recently read a paperback about a "Reluctant Messerschmitt." The ME-109 had caught many fishing nets and become the focus of an intensive recovery project. It was recovered from the channel near Hythe.

We are coming closer. Memories are more vivid, some sharpened by the Channel and the swastikas back at the Lashenden Air Warfare Museum.

Chapter 30:
The Landing 84

Dover, its white cliffs and hovercraft port, is a far different departure and crossing than forty years earlier. An announcement is made; the long discussed, vaguely initiated tunnel under the channel will be built.

Red, Chris, Don, and I, luggage and all, are the only cargo on the C-47 departing England heading across the channel to the beachhead in Normandy. I drowse briefly on the way. I believe I am a pretty cool character, napping while en route to combat. Cool, until years later I read both animals and humans sometimes are induced into slumber when they are scared. That is probably true when there is nothing you yourself can do to enhance your situation.

Over Omaha Beach, we look down on battle scenes of six weeks before. Both water and shore have witnessed battle carnage. Real, no speculation now. We are ready to learn. Apprehension and jumbo anticipation spur thoughts. "What's next?"

On the ground, a GI meets the plane. We are just "ho hum," four Second Lieutenants, four footlockers, four B-4 bags, four parachute bags, and four handbags; twenty items loaded into the back of a 6x6 truck. The C-47 is loading casualty stretchers for the return flight. They expected only a short wait. The St. Lo breakthrough was two days ago. The breakout is developing. The four of us with our sixteen pieces of baggage fill the whole back of the open army truck. We are wearing Class A uniforms, gas masks, and steel helmets. Silver wings and gold bars are our mark and emblems for accomplishments so far.

The message coming from the land reads like a month-old weekly. Shell and bomb shattered trees, churned-up earth, hedgerows, and burned out vehicles. We drive for some time on battered roads, their shoulders choked with battle debris. My sense of time and direction says, "Much farther will be too far."

We are volunteers, our camaraderie is continuous since preflight school at Santa Anna, CA. Chris is the only bachelor. Yesterday, a sinking feeling hit. Would we be separated? In the combat assignment meeting, the officer in charge said, "Sign yourself up on one of the group lists."

In a couple minutes, the four of us intersect in the middle of the room. I give a quick utterance about, "Let's keep together."

We disperse again to reconnoiter the lists. Pilot needing fighter units from both the Eighth and Ninth Air Forces were represented on clipboards held by disinterested personnel from the replacement unit. High profile groups are filled quickly. This scramble for assignment is unlike any procedure I have ever encountered in the military. Always before, you were ordered, assigned, or responded to a request for volunteers.

I think Red and Don found the clipboard with the opening; one needing four replacements in Normandy on the beachhead. We'd stick together. We signed on.

We are the first replacement pilots for the 377th after moving onto French ground. A tarpaper and wire mesh runway with dozed up revetments; P-47s, some with invasion stripes still in place, look more businesslike than training planes. This is the "road team." We wonder what is ahead. All four of us are assigned to one squadron. Still in the six by six, we hear a voice inside headquarters (a stone barn with a swinging wood door like the milk cow shed at home) say, "Take 'em over to the 377th. They lost four yesterday."

The truck drops us at squadron operations tent. A poker game is in progress on a blanket alongside the ops tent. Overhead, green apples are on the trees shading the organization. A well with rope winch and bucket is picturesque and off limits. The squadron's parachute tent is nearby. Lucash, squadron rigger, is about the first to give us attention.

"What's . . . ?"

The words broke off. Red, Don, Chris, and I hit the opening of a covered slit trench at the same time. Everyone else is already in. We hear incoming 88 shells "whoosh" ker-boom, explode. The ground shudders.

We are never last again. After all, we were on Luke Field's champion athletic team. We continue to exercise our prowess by digging deeper, sprinting faster and sleeping quicker in foxholes.

This time, Peggy and I sit up front on the starboard side of the Hovercraft. It vibrates and roars loud. We don't see a thing other than water and fog. It gives time for mental transition. We sit close, sharing thoughts, grateful to be here together.

George Taylor, courier and history buff, greets us outside customs in Calais. We also meet our talented driver, Hans. Hans maneuvers the coach, and George very cleverly maneuvers, guides, and governs us on our stirring adventure. The people we depend on are tops.

We load up under the first day strict regimentation of Hans and George. "No eating on the bus!" (Our responsible and cooperative nature soon relaxes our keepers.) George is custodian of the cold drinks and the microphone as well as guardian of the mike cord. We are on our way.

I have airfield A-12 and Omaha Beach on my mind, but we've quite a way to go first. Soon we are at Le Havre and crossing the Seine River. I enjoy the countryside as we speed along. Beautiful roads and fine fields, orchards, gardens, silage bunks, are interesting, especially to a farmer. Lunch is pretty much dependent on Chuck's French. Neither Peg nor I speak any. Lunch is in an upstairs restaurant in Abbeyville.

I remember Chris and my Normandy excursions, buying fresh eggs and going to the village café for veal and tomatoes. *Chris, I appreciated your skill with French people. We enjoyed the leisure after missions.*

After lunch, our group gathers at the coach, sort of beginning to settle into routine. George gives us historical information and answers many questions. Peg and I constantly watch the countryside unfolding. I keep my camera busy snapping farm, forest, and streams.

George enlightens us about the Bayeux Tapestries and the beautiful Bayeux Cathedral. Peg and I admire the magnificent tree in the cathedral yard. The tree, in my imagination, rivals the huge black walnuts Dad showed me when we visited the family home in North Carolina's Great Smokies. I remember. I was five.

The tapestries depict the battles of kings and knights and victims and victors. The animals and common people are shown in very explicit scenes of lusty living and dying. The story, nearly a thousand years old, unfolds on the embroidered tapestry two hundred and thirty-one feet long.

This has been a long day. We are pleasantly tired when we arrive at the Chateau du Molay, with its beautiful open grounds, ducks, geese, deer, a goat, and a friendly hunting cat.

Dinner is served for our group in the dining room with pink tablecloths and pink flowered china. Peggy tells me there were also potted pink begonias. Photographs of Bayeux Tapestry panels are on the walls.

Peg and I are up with the sun. We walk along the stream, visit with the deer and are followed by the friendly cat who doesn't understand English. It is quiet, no apprehension, just walking in Normandy with Peg. We take several pictures of us together using the tripod.

Filled with anticipation, we board the coach. Omaha Beach, A12, Mont Saint Michel, and Rennes are ahead. I feel the war vividly here. The years roll back. I remember the scenes as I saw them then. How beautiful it is now. The apple trees are blooming. Down on the beach the tide is out. I had not walked up this beach. Some in our group crossed this sand after D-Day. Even so, we are in awe by what we see and feel. The breakers are nonexistent or far out. Only a half dozen people are visible. A family looks at tidal pools. The sand is smooth. It is probably six hundred feet to the water, with a very shallow slope. On the entire beach there is only one vehicle, an old farm tractor. It is hooked to a badly corroded boat trailer.

Thoughts are both muted and pounding; each of us with our own interpretation. This really isn't my personal spot as an individual, but it is a very meaningful place as an American; and a human being. I take pictures of those who have been here before. We join in thankful remembrance with a memorial service.

The vivid memory I have of Omaha Beach is of Hans, our driver. He speaks neither English nor I any German, but we seem to understand each other. Hans is very careful for us to know that he is Austrian. He appears to be in his late twenties to mid-thirties, is sharp and extremely conscientious. While we are clustered in little groups, talking and looking along the breakwater, I look for Hans. He is standing, tall, blond and quiet, looking across the sand of the long sloping beach towards

Dover.

Chris spoke both French and German; was talking to flight mates over Alsace Lorraine where he died, was buried in France, later brought home.

I have memories without rancor and with gratitude and appreciation for what is ours. Somehow, the future should be better for what we have done. We who survive should be a part of that betterment. Omaha Beach is beautiful, a little quieting and serene; we laugh a little and chat. Those landing and walking from here are remembering.

Not that way with the Normandy American Cemetery. It is beautiful, somber, and startling. The quietness is in tumult. The rows of crosses are in perfect symmetry, but one's thoughts are not tranquil or ordered. Engines roar, guns pound, shells burst, bombs shake the earth and all the players are young. As a spectator today, I am subdued and thankful. I am wondering. Why?

With one exception, we will not see any further memorials of the Second World War until Luxembourg.

For the most part, in France, World War II seems to be just a happening after the 1914 to 1918 war. Every village has a monument to World War I. World War II is not the bleeding of France that World War I was. I'm not sure there was enough remaining to bleed.

Chapter 31:
Combat

Hans drives across cobblestones to Balleroy's square and city hall. Our leaders talk. Peg and I stroll to an obelisk in memory of area men killed in World War I. Peg's great-grandfather's family name, LeFevre, was one of names listed. Her ancestor was a French Canadian from Montreal.

"Headquarters" says, "Up the hill to Lignerolles and Romesnil Farm."

"There's the drive, the orchard! It is here. The field where the horse grazed."

I wrote home about the horse. I sat on his back the day before he was killed. German anti-personnel bombs aimed at our bivouac orchard mostly overshot into the field, killing the horse. One bomb hit on the left fender of a jeep parked with the passenger side next to our tent. Memories; fox holes, tents under apple trees, incoming shells, steel helmets and the nightly air craft attacks. I still feel conspicuous remembering looking up from our foxhole, seeing parachute flares lighting the whole area.

Down a short lane by the orchard stands the "Chateau Des Mogin's Maulers." Now, the chateau fronts a paved farmyard behind a fieldstone fence. The shops and equipment are on the left side of the yard, the barns and livestock are opposite. The farm looks well kept, a good operation. A tractor is broke down, parts spread on the stone paving, and a repair man is working. Both rear wheels are off; gear case is exposed suggesting a major power-train problem. My rancher reaction is a shiver for the down-time and a shudder for the final repair bill.

Forty years back, beyond a small field, ground crews repaired our "T-Bolts." Squadron commanders put second day "foxhole veterans" into the air. We were needed. Back then, the land was being churned by tank treads, bombs, shells, dozers, shovels, and combat boots.

Today, the owners invite us into the chateau. The stairway off the entrance is as I remember; so is the alcove under the stairs where pilots cashed mission chits for the two-ounce liquor ration per mission. I remember cold pancakes and cold syrup on tin plates served in a breakfast chow line.

I felt sorry for the mess sergeant who rustled some fresh cow's milk. He was so pleased to offer the treat at the chateau. I cringed when Colonel Magoffin said we couldn't use it because of the chance that it came from untested cows. Group Commander Col. Magoffin's statement was neutral and direct. The soldier's face alight with anticipation changed to dejection. I understood both.

I know the mechanics of command and assignment much differently after responsibility for operation of our ranch. I have been rewarded by several talented young men who worked for us during high school and college years, two of whom, Scott Wheelhouse, seventeen, and Ron Thomas, twenty-one, are presently doing the day to day operation while we are away in this pre-harvest ranch time, 1984.

Eager, I walk to the apple orchard. I remember a picture taken forty years previous; find the spot where Red, Chris, Don and I dug in by relating to the trees. The trunk of one pictured tree, recently severely pruned, still bears a long folded slash mark and has a distinctive lean. It was a few feet from our tent. In the picture, my buddies are heating water in a helmet in order to shave. The orchard is still a cow pasture, has an electric fence around it with a few small thistles along the fence and paths used by the cows. The original pictures have added worth now. You'd never know we'd been there. I've found arrow heads in the sand on Shutler Flat. I wonder what will be uncovered and not totally understood in this orchard at some future date. Certainly these filled-in foxholes contain artifacts of American youth in battle bivouac.

Red from Newton, Iowa; Chris from Modesto, California; Don from Kodiak, Alaska; and I; all had land and open space backgrounds. In the pictures, we are still young, well trained, inexperienced, all living. Perhaps our camaraderie then, gives meaning to the years that follow, for the ageing three of us, and the loss of Chris so soon.

I sit on a downed apple tree trunk. For a moment, we are again all young and present. I listen, swallow. I even grin. I stand and turn around: The roar is quieting a little. The view is not quite clear; the lump in my throat eases a bit. I look back toward the "Chateau des Mogin's Maulers" then step carefully through the cow pasture and wave to Peggy, glad for our lives. I have been here before. I am not finished yet. I have things to do and the support to do them. I am glad to have come back and maybe a little thankful there isn't more time. I set my camera and take my picture. Peg says George walked by as she waited for me in my reverie. Peg picked up a piece of the apple tree's bark.

Ken Wallace was wounded by shrapnel there amongst the apple trees in a bombing raid. He tells us, "I was in my foxhole. It penetrated my side and stopped against my stomach lining." The 378th flight surgeon took him to an Army field hospital, where they probed and retrieved the small metal chunk; kept him for three days.

A-12 is the beginning of my active combat war. We are near the front. Our planes are in battle when still in the traffic pattern. We can hear German artillery. We are bombed at night, shelled day and night.

The first two nights, we jump from our army cots under the side of the tent, into the foxhole, three or four times a night. No matter who is first, he doesn't stay on the bottom very long when bombs are dropping. It seems desirable to be on the bottom of the heap with your helmet on and face in the dirt. The bottom guys are continually displaced to the top of the heap by burrowing buddies. This can be described as either "foxhole rotation," or the "scared pilot cycle," each feeling pinpointed under the glare of enemy parachute flares. Our scramble seems involuntary to make contact with the earth on the bottom of the pile.

Frogs have the ability to elbow deeper into the earth. My dictionary also defines "frog" as being an offensive term for Frenchman. I remember the term being used in the World War I "Battle Aces" dime novels (nickel cheaper than "Flying Aces"). If "frog" for Frenchman had anything to do with being able to melt into the earth and live in the muddy battle fields of the trench-war and no-man's land of 1918, then I think it probably started out as a term of envy or admiration.

The 362nd Fighter Group had been combat active at A-12, Normandy, only a few days when we join the Group. Until July 20th,

362nd planes flew combat from Maidstone, England, across the channel. Even though they seem old-timers to us, they were newcomers eight days before.

Mud and grim war is fairly new to all of us. Diving into a foxhole several times a night isn't conducive to rest.

First thing, second morning, we start looking at various diggings. After missions today, we start digging four holes within our tent, one for each cot. We dig down about three feet and deeper for bunks' legs. We cover the head half with logs and dirt.

While we are digging, several of the "old-timers," week before us, come by poking fun. They say, "Jerry is being pushed back, don't worry." The real "old-timers," the ground echelon, those who have been here three weeks, whose foxholes we copy, don't say much.

Next night, enemy planes hit our orchard and blanket the pasture next door with anti-personnel bombs, kill the horse, wound Ken Wallace. Parachute flares light the place up like a stadium. Even under logs in my 'bunker' under our tent, I feel like a fish in a rain barrel.

I don't feel that way in the cockpit. Respect, yes, but not the same kind of fear. More like football. The objective is paramount. In football, skills that prevent broken legs and faces come from instinct and training. The objective is the play. In a fighter plane, be aggressive-smart and be sharp. As Colonel Joe Laughlin says, "We were scared but not afraid."

Next morning, we wake to the sound of "chunk-whoosh, thump-thump." We stick our heads above our burrows. Shovelfuls of dirt are flinging all around. Some digs are outdoing us. No longer are we the unnecessarily shaky newcomers. Several tents were pierced by anti-personnel, butterfly bomb shrapnel. One pilot's uniform hanging on a hanger has a piece of bomb fragment right through just below his wings. Down at the Operations tent, Harry Kraft's metal name strip on the ready board is bent out by a slug-sized fragment coming through the plywood.

We don't have long to wait before flying. We are sent to the tower, a shack alongside the runway. We are told, "Watch takeoffs and landings and listen to radio procedure."

The next day our names are on the mission-board. We are flying.

The squadron needs pilots. They expect us to be able to fly the airplanes. We are indoctrinated between the time of tail-wheel lock at start and wheels-up at lift off. We can be over enemy lines within seconds, one and a half miles. Actually, we are in range of enemy artillery and anti-aircraft guns on takeoff. We make right hand pattern taking off one direction and left hand pattern taking off the other.

After missions, Chris and I walk through the countryside. We buy tomatoes and eggs. Chris talks daily with Ramon, a French youngster about ten. I snap a photo of Ramon sitting in one of our jeeps. Close by is a well with a little roof and a winch and rope.

The chapel is in a shed attached to the cow barn. Our chaplain is Lutheran, Chris's church. We attend services and communion together. Seating is on bomb cases. Music is a very small pump organ. We milked our cow in a similar barn at home. The room isn't much different in size or austerity from our little country church.

It isn't a worrying time. We want and get lots of missions. We are learning. The missions are eventful. I fly John Hill's wing several times. He never fails to find good targets.

I am fortunate to fly Frank Pepper's ship on lots of missions. He instructs me to take care of the airplane. "Keep the RPM up. Don't burn up your gun barrels. Touch the trigger; keep the guns cool."

I can understand him and his consideration for equipment. Not much different than Dad's, "Pull down on that rabbit, don't waste shots," or Dad's "Shift gears, don't lug the engine."

I learned both at age ten. It was plain horse sense. "Take care of 'em so they're working when you need 'em."

Frank was later selected as best fighter pilot "bridge buster" in the 9th Air Force.

We are at A-12 for only a short time, two weeks, but it was here the four of us felt accepted. We were each raised "on-the-land." I think it helps us adapt to the beachhead, rough it, circumstances.

Marty Lucash, parachute rigger, sewed our parachutes' harness straps in place for a very snug, custom fit. I wear buckled and laced combat boots. "Chute won't break your back and you won't be left barefooted from a high-speed bailout."

My developing philosophy is, "Take care of the things you can control." For all the rest, "Be a team player, alert and decisive."

At A-12, I begin thinking a prayer and sensing a "Peggy-kiss" behind the ear simultaneously with locking the tail-wheel on takeoff. At A-12, I begin practicing quick exits as soon as I cut the engine on

Combat

parking. Like football, hunting, bucking wheat sacks, running a mile race; it just seems right that I have a plan and to sharpen it with practice.

We like to look whetted and keen coming in for landings. Want to let the ground crews know we appreciate their skilled work; to say we have pride in flying their coddled craft. Coming in after missions is designed to precisely get units on the ground quickly, avoid surprise attack, and to protect the runway, the aircraft, and people. All pilots, from CO down to tail-end-Charley, want landing formation to make a statement. It is a thrill.

Close, straight in, flight of four ship echelon right, accelerating shallow dive, low over the touchdown spot; sharp peel up to the left in four circles shaped like hoops on a tilted vinegar barrel; wheels snap open and lock, flaps down after the apex. The runway spacing will be leader turning off, two rolling, and one touching down.

Sometimes, it is wait your turn for belly landing, or blown tire skids and tumbles. Landing is an emotional time, a mixture of feelings, at times with information, at times with questions, elation, disbelief. We newcomers don't dwell on price at first. We have yet to pay. Debriefing, number of rounds fired, the condition of the guns, holes in the plane – all bring messages back from a mission.

Plane missing is different. The message is telegraphed by the formation. The eventual story can get better or get worse. Combat ground-attack flying is a mixture of skill, talent, and luck. The final score likely depends on unknowns.

I am completely confident in our armament and the airplanes. Our ships perform beautifully. As a replacement, learning at A-12, I mentally salute all predecessors both on the ground and in the air; want to meet the requirements.

May 1984: We are driving down the road looking for the location of the airstrip. "There,"--on a little rise. We drive a little further and turn around. A dairyman working in his farm lot alongside the road remembers and points to where the runway was. He converses with those speaking French, Chuck and George. He signals to follow him around the building, smiles under his cap and points to a steel mesh section from our old air-strip. He is using some of the runway mesh to fence their flower garden.

I have rapport with him as a farmer; doesn't seem too different than a neighbor. Ruddy-faced, a little round, not too cosmopolitan. He wears sixteen-inch dairyman's boots, but when it comes to soil and livestock, sun and breeze, heat and drought, and tenacity, I'll bet he has the "right stuff" (a term used by my father); I think it came from Dad's childhood home in North Carolina's western mountains. I'm glad Chuck Yeager used the phrase. You'll know it when you see it. I saw the "right stuff" a lot in combat and afterward.

I write to Peggy daily and to my father a couple times a week. In an early letter to Dad, I write:

"Missions aren't hard to fly; they just mean keeping your wits about you and keeping your head on a swivel all the time. We are not allowed to write about missions or any encounters with the enemy. So those bits of information will have to wait. How are your chickens doing? I could sure use some fried eggs for breakfast."

A week later I wrote:

"I've been averaging about one combat mission a day lately, though I haven't many missions yet. This combat flying is just like tending header in short grain. You've got to stay awake all the time to see that you don't pick up any rocks. I've developed a screech owl neck and intend to bring it home in one piece."

Both letters were inspected and passed by a censor, uncut. A screech owl in the open Columbia Basin country of Oregon is a small ground-burrowing and fencepost sitting little owl that seems to swivel his head on his neck a full 360 degrees. Tending header in grain-harvest in the short grain years of the thirties was a job that took all kinds of concentration. If you missed seeing a rock, while cutting close to the ground, there was hell to pay. The results were loud and destructive as it forced its way into the separator's cylinder.

The sum of one's young lifetime of experience comes into play when pilots go "automatic" in combat situations. If I were to describe only one of those first missions to my Dad, it would be the fifth.

We are a two-ship flight; yellow leader and I his wingman. We are flying in battle formation; meaning I am out a ways and nearly abreast of the leader; a man I admire on the ground and am coming to the same conclusion in the air. John is a squadron original pilot. I size him and maybe three others as tops in our squadron.

We are flying at about a thousand feet above the ground under an overcast sky among pillars and smudges of battlefield smoke and dust. We are on radio frequency of an advancing armored unit battling through a flexible front. They call our code name saying they are under attack by two ME-109s. Yellow leader instantly turns, seems to know exactly the direction. I hang in position with my eyes sweeping high and low, trying to spot them. Yellow Leader answers my unasked question; jettisons his bombs, so do I.

Times as a youngster, I hear a mother hen's squawk in the baby chick yard, I grab my 20 gauge from the side porch wall, intercept and blast the hungry raider.

I spot one enemy fighter coming from the haze around a pillar of smoke. I call him out at one o'clock low. Yellow leader's answer is to break to intercept the 109. I feel elation and scan around for another enemy. The written and "pounded in" rule of the two ship fighter element is--the leader is the guns--wingman prevents surprise attack. I look and hope for another. I eyeball around in a wide, three dimension sweep. I swing back. John's 109's guns light up and tracers fling more in my direction but seem to be in desperation. John is closing; I do another visual sweep to cover his pursuit and join just as his tracers converge into the root of the 109's left wing. The enemy fighter plunges into the ground. The other 109 never shows.

Those moments alone in the apple orchard that day in 1984 can never be relived. The feeling is unique. I reminisce through my catalogue of life; probably every day since that long ago time has been influenced by those surroundings.

We eat lunch back in Balleroy. We learn the difference between "small shrimp" and "small snails." Seems some of Peggy's advisors don't know how petite their French skill is. She isn't expecting miniature snails. She is game and glad to share. She saves the eating tools, a pin and small cork.

This time, I see Mont Saint Michel from the ground. After a day, which includes visits to Omaha Beach, the Normandy American

Cemetery, and A-12, it is a tribute to Mont Saint Michel that we have emotion enough left to appreciate its lore. From the cannon and walls of the first level, through the shops and sanctuary zones on up the narrow steep pathway, I feel steeped in history.

Peg and I are inquisitive all the way to its top. We pause in a chapel on the summit for a moment of meditation. Out in the bay is another smaller rock, an island not too far away. My log book notes 'island' in missions listings. I look down from the heights through an ancient firing-slot.

I remember an earlier time of my own, the sensation of a wing over to a vertical steep dive actually having to push the target into view from a near over run on my first dive bombing missions. I triggered a 50 caliber burst in response to tracers rising to meet us.

We learned fast, the book on the P-47 and its missions was being rewritten during our first three weeks in combat from French air strips. Like the muskets of the *seventy-sixers*, the P-47 responded to the can-do headset of its handlers.

The need for close support for allied ground troops was obvious. The tactic necessary was just as obvious to skilled adaptive leaders. The beachhead pilots of the Ninth Air Force lived on the ground with the dust and grime and constant enemy harassment. It wasn't necessary to explain to us the need for our developing missions more than once. We could save allied lives and be the difference between stagnation and movement, stalemate and liberation.

The T-bolt's tremendous pilot protecting and bring-em-back record added to the confidence of the pilots. The troops on the ground loved the ship.

Chapter 32:
Rennes to Paris

Rennes A-27, a former German established base in Brittany, had bomb shelters, barracks, hangers, runways. When we flew in, some of each was still unimpaired. Buildings and bomb shelters concealed trip wires and booby traps.

The ground war mainly fanned out towards Paris and along the Loire River to the east. We shared close support for troops with other groups in bottling-up Brest harbor and its sub-pens. A real slug-it-out clash hammered on the ground surrounding Brest. Fixed batteries of German 20mm, 40mm, and 88mm anti-aircraft guns and our developing techniques foreshadowed Rhineland battles against hardened defenses.

One incident: A company commander called for support for his troops. He asked for bomb hits on, "--the other side of a wall." His urgent words sounded his predicament. He needed help. We obliged. His appreciation showed clearly in his tone after we hit. We paralleled the wall, didn't overfly his troops.

We flew missions, swam in a rock quarry, and dined downtown. We slept above ground in buildings at A-27.

In previous wars, not many have been able to return to battle sites. Better transportation and communication facilities, longer lives, and more leisure time, enable us to return and look back where an important part of our lives took place. Perhaps returning and recalling will make a difference.

We drive around the airfield and stop at the far end of the runway. It is raining. Mack and I flew from this runway. Peg and I walk a short way. The edges are overgrown with brush. I pick up a "memory-rock."

Back on the bus, Paul leads us in a memorial service; there in the drizzle, by the railroad tracks, where we had stood before.

French citizens walked out near this end of the runway and watched as we took off. Their Sunday strolls to the airdrome's perimeter seemed to be entertaining for them and us. I remember what first appeared to me, to be a distinct difference between French and American communities. They, like us, took Sunday off and like my community in the summer after church, spent Sunday afternoon relaxing or visiting. Unlike us, they also dressed up and seemed to do similarly on Mondays. Their Mondays seem to fit more with our actions on Saturday. Like us, they went swimming or viewing the sights depending on age and agility.

Eastern Oregon farm and rangeland local customs differ from the French. After a Sunday ground support mission to Brest, a jeepload of us were given the OK to drive a couple miles to a rock quarry pond for a swim (Doc May had checked the water). We relaxed, breezing down a narrow tree-lined road; the wheels even tossed a little gravel on the corners. Rounding a sharp turn, we met a foursome about our parent's age, nicely dressed, the two ladies in the lead, the men five or six steps behind. One gentleman was standing on the road's narrow edge, facing away from the ladies, as we abruptly rounded the corner. The surprise was complete for all parties. It is forever framed in my mind's eye. The ladies' mirthful expressions showed little sympathy as they turned, checking the busy victim's plight. Both of the gentlemen seemed relieved when we sped on.

Another day, Chris and I joined a sightseeing group of civilians at this railroad end of the runway, watching our fellow compatriots take off. Chris's French was passable. We watched as each loaded pair of Thunderbolts lifted off. Our eyes followed each element until wheels up; then they visually swept back down the runway to the next twosome, already rolling. On one return sweep, I glimpse head and shoulders of a young adult beyond a low stack of lumber. Her eyes and my eyes didn't meet. Certainly I would have been the one embarrassed.

We adapted to France's street side facilities for men. We attempted to achieve the proper composure. After all, hadn't we former aviation cadets, almost overnight, achieved the newly winged Second

Rennes to Paris 235

Lieutenant's version of the tilt of the thousand-hour crush to our officer's hats? In those days, we were adaptable.

Come to think about it, we still are. This wonderful group of twenty-year-olds of yore, now forty years later, is again adjusting to necessity as we tour rural France. We are still adaptable.

I find the thick walled German bomb shelter. It is partially hidden by newer buildings. Four decades ago, it stood in off limits isolation about forty yards from squadron operations. Keying on it, I discover the house where we were quartered. I take pictures. The entrance has an identifiable arrangement of bricks in stucco. I have a photo taken by Don Clark of eight of us standing in the entrance in variations of grinning nonchalance. All eight survived the action of World War II; certainly exceptional if not unique for a group photo of eight replacement fighter-bomber pilots.

Some of our original pilots were rotated home for a month's leave from Rennes. I was in the ready room when Doc Mays told Harry Kraft, "Grab your things, you're going home."

The C-47 was almost ready to depart for England. Harry scurried. He came back to the squadron at Reims and flew my wing on his first second-hitch mission.

At daybreak one morning, a Ju-52 rolled its wheels down our runway before recognizing its error. The German tri-motor roared out and away going west. Could it have thought it was on the airstrip at Brest harbor? There was a slight fog on the runway and the anti-aircraft guns were caught not firing from a previous hold order.

Also at Rennes, it is put to me this way: "We are moving to Reims in a few days. Those with planes will fly them. The rest will ride the baggage trucks. However, we need to get the L-4 Cub up to the A-79. Alan Campbell, formerly Squadron now Group Intelligence Officer, needs a ride to Paris and someone needs to be at Reims to run the flare gun, and be landing officer when we arrive."

So, rather than being part of the baggage, I grabbed the chance. Alan Campbell was being assigned to SHAEF, Supreme Headquarters Allied Expeditionary Forces, in Paris. What a neat trip: Rennes, Le Mans, Chartres, and Paris. I think they were looking for a farm kid who could fly and land with a chance of getting the unarmed Cub safely to the next base east of Paris.

Bill Fisher checks me out in the L-4 Cub. The taxiway we used for departure is still used. Bill could be classed as a truly hot pilot. He is finishing his tour. I am just getting going. Bill has more flying time in P-47s than I have total flight time. But, I have read Richard Halliburton's "Flying Carpet" and World War I "dime novel" flying stories, "Battle Aces" and "Flying Aces." I had belonged to the Richfield Oil Company's "Jimmy Allen's Flying Club" when I was ten, and if that isn't enough, I am a Second Lieutenant combat pilot with twenty-six missions and next in line to get a ship of my own. I would call it "Sunny."

My checkout ride starts with me in the front seat riding, Fisher in the back seat flying. He starts down the perimeter taxiway, increases speed to lift off, turns 90 degrees and bee-lines over a ridge to an army field hospital, buzzes some tents, pulls up over an electric transmission line, finishes a 180 reverse chandelle, cuts the throttle, zips under the power line, over a ditch, and sets down in front of the buzzed area. I had received a few hours dual in J-3s before preflight school; just enough to weed out those few non-flyers missed in the sign-up screenings. We were not allowed to solo.

I know enough to be observant, but I really don't feel I am an L-4 solo pilot. I figured so far, it is for show. I will get the check out on the way back to the airdrome. Bill crawls out as he cuts the engine. Striding away, he gives me an over-the-shoulder, "Take some of the troops for rides. I'll be back in awhile."

The plane is already surrounded by about seventy-five or eighty "troops," some clamoring for a ride. Being soft of heart and on the spot and knowing Bill a little, I am pretty sure the ship will get out of here OK. Also grabbing for a bright idea, I come up with the obvious. Loudly, I ask, "Would it make any difference if you knew I have never taken off solo in this type plane?"

Sure is an emergency good call, because it thins the eager group down to a manageable three people and none of them very big. One knows how to prop a plane. He spins the prop and jumps in the back. I wheel around and taxi over the turf back to the edge of the swale, a little this side of the power line. I check the mags, make a power sweeping turn to align and gain speed; give a firm feed on the throttle. We head for a gap in the tree-lined hedgerow. We make it with inches to spare. It is probably my most harrowing takeoff.

We skim around for a few minutes. I come in between the trees opposite the spot I had taken off and roll to a stop. After my second flight, about fifty men are waiting for a turn. My frustration of selection is again made easy. Good old Bill has sent a First Lieutenant Army nurse out for a flight around the area. She climbs in and off we go. I have navigated in combat, located enemy emplacements, and spotted a low flying ME 109 in dust and smoke in response to a ground call, but I get lost with a nurse in a Cub. It is a couple frustrating minutes of unshared confusion until I spot the power line and follow it to the hospital's pasture.

The nurse is still thanking me for the flight when Bill trots out, climbs in the back seat and shakes the stick, signaling he will fly. He wheels around, revs up and releases the brakes. We take off the opposite direction to our arrival, into the dip, hop the ditch, under the power line, and we are off. This lesson qualifies me for my own hedge-hopping magic sojourn over France, to Paris and then on to Reims, the scene of World War I pulp magazine marvels.

The little airplane is heavily loaded because of all Campbell's stuff. I carry only a musette bag and camera. We land for some information at Le Mans. There, an L-5 pilot tells me they were flying into Paris but are waiting until the weather clears. Campbell and I elect to go on. We follow the railroad track, circle the cathedral at Chartres (I took a couple pictures) and fly on to Paris in deteriorating weather.

Today, we drive the road by Le Mans and Chartres same as Campbell and I flew forty years before. Hans stops and fuels the bus at a service station in the open country. Friendly black and white cows in the adjoining pasture look to be Ayrshire and Holstein crosses. They respond to my Eastern Oregon cow call. It works. They turn at the first low "whoop." I don't want to cause an incident; I quit calling after three or four start my way.

This time instead of circling, we stop and visit the Chartres Cathedral. George points out the difference in architecture of its two spires. I understand better now the land's resources required for constructing these huge cathedrals. Men and materials were gathered from large areas, over long periods of time, even extending into other countries. Both human and material resources must have been exhausted in the endeavor. No electricity, no phones, no fans, no refrigeration, no diesel; just wood, stone, mortar, iron, brass, lead, glass; rope, tackle, hand tools, muscle of both men and animals; patience, planning, perseverance; cunning, harshness, grandeur, inspiration, vision; love, craft, skill, talent, and gift: So very much must have been extracted from a region and swept into the construction and reconstruction of cathedrals. Can we fight and build and create at the same time? Or must we? It depends.

The weather is stormy when we arrive in Paris. We drive over bridged railroad yards, alongside canals, and bridges. Slower, but not much lower than in the Cub.

Alan Campbell is so happy to spot the Renault factory parking lot with the L-5s parked on it and most happy for the landing. He indicates relief from considerable concern. I am thinking, "Here is Paris. I have the equivalent of about five dollars, and I am wearing khakis and my hotshot leather jacket. I am also hungry!"

I shouldn't have given the situation a second thought. Poverty was not a problem.

Rain is falling now as then; Hans is a superior driver, narrow streets and all. The Eiffel Tower looms high.

In '44, I circle it twice while Campbell looks. I check for room for a landing roll, maybe stopping under the tower. Bill Fisher has shown me how. Did our squadron "wheels" know they could count on Childs? Is this their idea of punishment or reward? My enthusiasm for Richard Haliburton's escapades isn't in my 201 file. Only another ten minutes of descending rain and darkness will give me reason to plunk down at Tower Eiffel's base. What a "grandchildren" story! "Mission to deliver former squadron, now Group Intelligence officer, to Paris; landing in the fore bay of the Eiffel tower." I'll never know. Campbell brings me out of my fantasy, with his, "There it is!"

Rennes to Paris

I am ten minutes too early or ten days too late to pull off the landing in front of the Eiffel Tower. Paris has already been liberated. My reaction to the consequences from "higher ups" stops me. It wouldn't have been dangerous, not even daring; surely not too smart.

Next morning, I do a 360 climbing turn up out of the Renault factory parking lot and head for the new strip at Reims.

We of '84' arrive at Hotel Bergere. Soon we are on our way to a "left bank" bistro located down narrow stairs, underground. (Recalls memories of the Apaches' Dance at the Sigma Chi house, Corvallis.) Dinner, authentic native Parisian, is with the MacLeans and their daughter and son-in-law. He is French. They live in Paris. The proprietor is a friend.

Calvados does not taste any better now than forty years before. The quality product is Big Mac's treat, and it takes all eighteen of us but we down it. The waiter does a masterful job of dividing the Calvados dividend. I will never declare another.

The cafe is not quite exclusively the 362nd's. A young French couple is sitting opposite each other at a table along the back brick wall. Peggy and I are cozy, sitting side by side in a foursome at our table. From our vantage point, neither of us misses a most artful, only slightly subtle, set of creative foot and knee maneuvers. The couple is oblivious, but not us. We regard the demonstration as interesting and, at times, informative. Nice accompaniment to our chocolate mousse, specially requested by Big Mac.

However, the real thrill and outstanding events of the evening are, first, trying to hale a series of cabs for the return to Hotel Bergere, and second, the wild ride following. Paris law limits three passengers to a taxi. Six cabs in intermittent trail, remind me of a shot-up squadron finding its way back from a tough ground mission, on the deck, out of ammo, and semi-lost. Eighty kilometers an hour, seem like a hundred through narrow streets, often two wheels on the sidewalk, seldom four on the ground, make combat seem a little mild. Instead of tracers, we suffer intercepting traffic and blaring horns. We had stragglers then, too, but tonight everyone returns.

Next morning, we have a slight delay; couldn't have been more fortunate because Vicky, located by our fabulous leaders and Hans together, pinch-hit for a missing tour guide. George, though qualified, isn't permitted to guide us in Paris. Vicky, with spicy French accent, is great to listen to and see. I take notes with my 35mm Pentax, good for verification.

We tour from Tower Eiffel to the Water Cannons and then Montmartre. I look, marvel, and exaggerate with my memories. We men buy berets, each a different color. Later, we learn the local significance for some of the colors. Chuck's is blue, mine earthy, Paul's red, and Nat's black. We find "crimson red" is not considered appropriate for acting chaplain, Paul.

Before leaving us for her scheduled afternoon tour, Vicky suggests we change our Sunday plans slightly and tour Versailles this afternoon instead of tomorrow, especially since Versailles is closed Monday!

I did not know Versailles is nearly four hundred years old. The fountains were first watered and pressured with the Marly Machine, a system of fourteen paddlewheels, each about 30 feet in diameter, turned by Seine flow to power more than 200 pumps, forcing river water up a series of pipes to the Louveciennes aqueduct, a 500 foot vertical rise. It was in use from its completion in 1688 until 1817.

Versailles's cobblestones, fountains, statues, trees, and people are fascinating. Light rain is falling. Contesting acrobatic aircraft are competing toward the west against a "Monet" sky. I take pictures. Peg tells me I am probably the only visitor to have pictures of the surface drainage systems in the Gardens of Versailles. Some of us are farmers; makes life better for all doesn't it?

Next afternoon, Peg and I stroll window shopping along the Champs Elysees. We have lunch in the Renault show room. A highlight of Paris for us is having coffee at the Raphael Hotel. It is very elegant. The lobby is done in dark wood burnished to a gloss.

I had been here forty years before. Alan Campbell, Major Jeffery Gordon-Creed III, and I shared a suite of rooms. Gordon-Creed is in British battle dress with a backpack, arriving from Crete and the underground by way of England. He and Campbell pool resources at the desk and gain lodging at this super brass inhabited Hotel. I ask Campbell,

Rennes to Paris

"Just how elegant is this place?" Campbell had lived nearby during his Paris years. His reply is simple. "Ali and Rita stay here when they're in town." (Ali Kahn /Rita Hayworth)

I am convinced the accommodations will be adequate. Silk wall paper, mirrors everywhere, sitting room, two bedrooms, two baths; cot in the living room for Second Lieutenant Childs: Quite a change from a foxhole; or even the two-story former German quarters at Rennes.

Is this what wars are about? We'd never have been here if Alan Campbell hadn't known his way around. The military had taken over the hotel. Someone hadn't chinked the door so I, too, "Campbell's pilot," enjoy the "high on the hog" scene.

Campbell leads us to his old hangout below the street, the American Bar. Johnny, the proprietor, greets him with a hug and a smack on each cheek. He spotted Alan before he reached the bottom of the stairs. I ask the pianist if she will play St Louis Blues. There goes a fourth of my cash, but the song brings me closer to my wealth, Peggy. The keys form the tones; my mind the picture.

We share champagne in a Brigadier's party. I have a French phrase book with which the General busies himself while Campbell and Gordon-Creed engage his date in lively and most animated discussion. Since I am near broke and completely out of my element, I think the least I can do is enjoy myself. I collect the cork from a bottle of the General's champagne. We follow Johnny's suggestion for dinner. I pungle up my entire funds as anti for the endeavor. It is a short walk to the selected restaurant. The air is nighttime in Paris (hasn't changed a lot).

We dine and wine in elegant surroundings. Gordon-Creed suggests we look up his aunt who lives in Paris. He, however, reconsiders, saying, "She probably has her bloody head shaved as a collaborator." and "It isn't too good of an idea."

Wanting to carry my weight, light in the pocket as it is, I suggest it would be nice if we could find another officer of high rank to endow us with a round of champagne before we turn in. I am catching on fast. Raising an eyebrow, I address Major Gordon-Creed with, "By the by, old chap, what kind of an officer is at the table with the beautiful platinum blonde?"

He responds with "Oh, I say, I do believe that is a bloody Admiral in the British Navy."

"Do you suppose you could get him to spring for a round?" say I.

Jeffery says, "I'll do it. I'll tell him I have a couple of junior officers here from the colonies who believe it would be beneath the dignity of a senior British officer to buy them a drink."

Up until now, I am enjoying this very much. However, Jeffery makes his move in response to the Admiral's preparation to depart. Gordon-Creed is super. I watch not believing my eyes when the three of them start walking toward Campbell and me. The Admiral, the first in Le Havre harbor, is extremely kind and gracious as he joins us.

Again, Campbell and Gordon-Creed verbally descend on the lady. I talk with the astute Admiral. I speak Western stammer mixed with "Box La Rue" French. The cork of the Admiral's champagne indicated vintage 1927, the General's, 1937. I keep the corks.

Next morning, as I come out of the bath, Jeffery greets me with, "I've a bit of a two-two here with a silencer. I'm for potting some bloody old pigeons."

He takes a .22 caliber long-barreled hand gun from his pack; holds it with both hands, steadies his arm against the side of the open casement window and plinks four in succession off the sloping gable's roof at least fifty feet across the street. Three birds are rolling without flutter at once. With this somewhat casual display, Jeffry asks, "I say old boy, care to have a go?"

I'd shot Canadian Honker wild geese from a horse at daybreak before breakfast, but I had not had champagne the night before. I'm glad I am not wearing any gunnery medals. Never before have I declined an opportunity to shoot. For breakfast, we have Spam under glass.

I wonder if Jeffery is living; it's a long time since. I hope he survived. His assured and polished manner, along with his personality, impressed me.

Alan Campbell sent greeting cards to Peg and me through the years, from London, New York, and his home in Pennsylvania.

Peg and I have a brief encounter with a persuasive, perhaps a con man. We leave the Raphael and step out on the sidewalk. Quickly, a car pulls up beside us and through his curbside window the man asks for street directions, saying he is a stranger in town: Then says he wants to

Rennes to Paris

give us some valuable looking new clothing stacked on the rear seat, leather jackets and that sort of thing, simply because he is going back to Italy and "You are so kind to answer my questions, so courteous to a stranger in a strange land."

Peg catches onto him immediately. It takes me a couple of seconds. I am still on my magic carpet of forty years ago, but we manage to evade.

We all attend the Follies Bergere with Hans as tour director. Regardless of the pictures Chuck takes of my observing "intensity," I remain cool and composed, extremely interested in the equipment and scrutinize as best I can the contraptions used to assure the safety of the Mademoiselles. I am also concerned they may get chilled so my applause is exceptionally warm throughout the whole show.

Peg and I join others in early morning shopping, and then walk to the Louvre. On the way, we stroll along the Seine and take pictures. Across the river, we can see the towers where Marie Antoinette had been held prior to her beheading. History tells us when too many ride, and too high, those being ground under finally resist and the system falters. The portrait paintings at Versailles of her and her children and husband, Louis the XVI, impress me. I remember those paintings later when George mentions we are in the village of Antoinette's capture and return for execution.

We appreciate Mona Lisa, the Winged Victory, and Venus de Milo in the Louvre, but Peg and I are especially caught up watching the excavation in the center courtyard. People are doing what seems to us to be an intensive dig and examination of the spoils coming from this huge excavation. Two backhoes loading two endless belts are being worked by examining crews. A former tower's base is exposed, apparently having been bound in silt for many centuries. We later learn this is to be the site of architect Pei's "glass pyramid," a new entrance to the museum.

We walk back to Notre Dame Cathedral where we thrill to the music. The choir is making an entrance for the eleven o'clock Sunday service.

Like combat leave forty years before, seeing Paris, its people and its sometimes light and sometimes very heavy grandeur, comes to an end. It is Monday morning and two little girls in yellow slickers with books and lunch pails are skipping up the sidewalk to school as we are loading on the coach. Having been out of large cities or even our own Portland for some time, watching people and their dress and actions is always interesting.

Chapter 33:
Dayton Ohio 1960:
Our Billy and Our General

Paris with Peg is quite wonderful. I am now anticipating the sojourn on to Reims. However, I have a nagging, self-inflicted dissatisfaction because Peggy was first to recognize the man encountered as we left the Raphael Hotel as someone with a dubious purpose. I excuse my slow uptake somewhat only because I was deep in remembering. Peggy had just snapped my photo standing beside the polished brass sign indicating the Hotel Raphael doorway. Our scene was a giveaway: American tourists. As I muse on the incident, I remember Billy King, our magnificent salesman. I grin to myself and relax. I wonder how Peg would do in a sale encounter with Billy. For those, if there is anyone who does not know Billy King, ask any 377th Fighter Squadron member. Billy is a sales force, a really capable, white-shoe influencer in "river-city" tradition.

The first 377th squadron reunion, 1960, is my first personal contact with my outfit other than with Chris's family and Don Clark. Peg and I decide she will stay with the ranch and kids. I will fly to Dayton, Ohio.

The DC 6 seems to me sure to end up in the river as we speed through the runway intersection landing in Denver. I had never experienced reversible pitch props. Flying out of Chicago for Dayton on the three-finned Lockheed Constellation is exciting for me but near the end of the prop age. This once magnificent aircraft is tired, its interior worn. Jets with high altitude flight are changing the whole scene even before I catch up to where we were. My last flying was in a Cub over the Vanport Flood in Portland, Oregon, June 1948.

By the time I register at the hotel, I feel more the country bumpkin, certainly not the suave fighter pilot of London and Paris. Not even the ex-POW at Rest and Recuperation in a Santa Monica beachfront hotel fifteen years earlier. I have been in my local beautiful world of family, farming, land conservation, improved schools, elk hunting and steelhead fishing. The first day in Dayton, I am not nearly up to the self-esteem of a graduating from preflight Cadet, nor even close to a Luke Field single engine pilot graduate with some hours in a P-40 War hawk. This is going to be something; uniting again with foxhole digging, flack-fighting survivors from our era of locomotive and tank busting. I have survived in prison camp and hospital; first under the control and whims of a devastated hungry enemy; then, for me, the deft challenge from a dedicated physical therapist for my legs. My face challenged the surgeons' skill. Dad's death before I was out of hospital; two months later, military medical retirement: Finishing school, rehabilitating the ranch, our good fortune with our children; Chris, born in 1951, and Kit, 1954.

The years have blended to diminish most of the unpleasant and life is good, however, I have not participated in comrade debriefing. I am apprehensive. My questions are: Will they remember me; why didn't anyone write to Peggy?

I got home before my belongings. I open my footlocker in the middle of Peg's family's living room. I wonder what message the comrades were sending to Peg when they packed my things: My letters, uniform, camera, photos, 201 file. The 201 file; what is it going to say? Did I help the team? I wonder how I stacked up. What was my grade? I gingerly start to read the assessment from Harding Field, Baton Rouge Replacement Training Unit.

Twenty-four training missions each with a grade of S for satisfactory or Good for better. There are no *Poors*. I relax a notch. I read, *S, S, Good, Good, S, S, S,* so much for the first 6 training missions.

7/ Good Join up; aerial camera gunnery, good; passes problem #1.

8/Bomber escort and offensive tac formation, good, steady formation, "stays in sharp break, good bomb hits."

My ego takes an upward nudge. I lay the file aside. What else will be in the footlocker? I had two quarts of untouched scotch (from mission chits) when I went down, I also had un-cashed mission chits, not here;

Dayton Ohio 1960: Our Billy and Our General

neither are my flying jacket, or my liberated appaloosa rabbit skin Luftwaffe jacket. My old leather brief case made out of tough bull hide with my personal indexed file is intact. So is my neck scarf of white parachute nylon. My bright-red lined London-tailored Eisenhower jacket and 1000-hour crusher hat remind me of the dress code of a cocky wood duck.

My salad days have been cut short. I wear dark glasses day and night for comfort and to not offend anyone. In recent years, two friends, both former POW mates have individually mentioned their difficulty in looking at my face in prison camp. One said, apologetically, "It sounds harsh, but I couldn't look at you."

When the footlocker is empty, I again pick up the 201 file. I have to know.

9/ Strafing and dive-bombing, good. Nice passes and good bomb hit, steady pilot, has confidence.

I am feeling better. I have been so dependent for so long. Even with my best efforts, I can't stand straight, nor can I see easily; my eyes are draining and patches of hide are raw. I have never complained. Others are much worse off than I. I want more.

10/ Low alt x/c (cross country) S, hangs around flight room, eager to fly. Yep!

11/Formation and combat, operational ceiling, Good. 2nd element leader, good form.

I read on down the page, line by line. It is a real boost to my spirits. At the bottom of the page it says *"over,"* and on the back, *"LT. Childs has proven to take to fighter planes exceptionally well and has prospects of being good material for any combat group."*

Maybe I was a net benefit after all. I didn't tell anyone or show anyone the training record, but it meant as much to me as the <u>posthumous</u> DFC. The DFC is very important to me. Peggy was so sweet when she said, "It came in the mail." The DFC is for hope. It belongs to Red Campbell, Peggy, and my Dad, L W. They earned it.

I find the squadron gathering room. Walk in. It has been thirteen years since the last surgery. I begin to recognize faces and remember names, the "Old Pilots" and the newer ones. I had flown forty-eight missions with these people and just left, sixteen years before, after an early lunch. No one speaks and no one nods in recognition. I am a little flustered.

I think I will try the offense. "I'm David Childs. I'm from Oregon."

No one responds. No real close flight mates are in the room but there are several I have flown with. Some have flown my wing. A few years later, one tells me, "Dave, we thought you were an imposter. Most of us knew you were dead."

I had been in contact with a couple of close friends in the squadron, but the word had not reached out.

Remembering Billy King started this bit of recall. This story is about his qualities as a salesman. When it was over, I felt I had been privileged to see one of the really great sellers of all time; con artist is possibly not too harsh, in action.

We are in Tom Beeson's room. Tom was a really talented squadron C. O. He gave his absolute best and his talent didn't come up short. When he was a boy, he surely reminded folks of the skinny long-necked, freckled kid in the "Our Gang" comedies. He was a top notch West Pointer. He found targets, was innovative, got results, and led the works. He was one of four "Old Pilots" I was especially excited to fly with because we would get things done on their missions.

On this 1960 occasion, General Tom is telling us about the suborbital flights of rockets. He relates a couple stories concerning the response of the monkeys in their training. He had recently been privileged to observe and be briefed on the program. He is in charge of the 105 program in the U S Air Force. I feel privileged to be in this old timers' bull session. I can sense Tom is going to go far. I hope the U. S. Armed Forces attract a lot of men as good as he.

Next day, he asks me if I would like a print of my combat film which is included on his film. Several brought film to Dayton. I say, "Not now, maybe later."

It was good gunnery and the pictures are clear. Shot the left wing off a ME 109. However, my kids are little and I think I will not bring the war that close to home for awhile.

It is a thrill to see my name and the date. I am leading an element. Tom is leading the squadron. He calls, "Two 109s, someone get the one on the left." I obliged.

It is important to know Tom Beeson is one of a very elite group of individuals. His training, ability, and sophistication made him a shoo-in to earn top responsibility. This bit of information is vital to my story.

The talk is sometimes serious and sometimes light. The losses to this small group of mostly original pilots have not the personal losses for me as for them. Mine are those in my combat time. The tragedy for me is seeing friends' names on casualty lists later.

We didn't seem to want to dwell on the past but more on now. The room's gathering includes some extraordinary talent. Impromptu and fortunate are descriptive words attachable to this occasion. Billy King, with a little encouragement, might have a go, others who have seen him in go mode, name some of his most famous spiels; "Oil Stock," "Health Machines" (vacuum cleaners) etc.

About this time, room service brings in a tray of ice and setups. While the waiter sets the tray down, a knowing someone catches Billy's alert eye. King kind of eases his verbal throttle ahead and his speech takes on laminar flow. His victim, the waiter, nods in rhythm with Billy's cadence. In seconds, the young man is nodding in agreement, even checks for his wallet before he knows what Billy is selling. No matter what Billy says, the mesmerized subject is nodding "yes."

I chuckle to myself and am sympathetic with the mark. It is awesome. Billy King is a virtuoso with his spiel. Quickly releasing his victim like a cow cutting contestant does when he's gotten his first critter bested and turns away and selects another from the herd and moves in. Before you could think FW 190, Billy has centered on Tom Beeson. I give thanks it is Tom, not I. What a showman! Billy has picked the sharpest and best of the lot. In ten seconds, Tom's lips relax. He nods in agreement as Billy turns up the pressure and verbally inundates the West Pointer. The General digs deep in his defenses, shakes his head and then goes on offense. He grins. Billy has gotten in a few hits. The script was in fun but I have seen the master of sales meet his match. The encounter is unforgettable.

Chapter 34:
Living

We stop near Chateau-Thierry Aisne-Marne World War I American Monument. The valley opens to the east behind the white vertical columns. This is my first opportunity to walk out through France's timber. Trees, possibly thirty or forty years old, are interspaced with stumps cut nearly flush with the ground. This is shell-shocked soil revived; it has tilth. There are no tell-tale rivulet tracks indicating water runoff.

We drive on through the beleaguered battlefields. George and Paul remind us with history about the people and their misery in World War I; foot soldiers slogged it out day after day after day; the terrible stalemates and the mud. They describe how miserable souls of one side, at one time, are so emotionally and physically gutted, they abandon their lines, and the equally miserable souls of the other side in like condition, didn't know. Or if they did, they are so drawn-down and weary, they cannot act.

We travel east down a two-lane road bordered by tall trees with branches nearly meeting overhead. Intermittent stumps on both sides of the road indicate the trees we are looking at are post World War II, not the ones hiding armored vehicles for us to ferret out and destroy. The current crop of trees are not shattered or scarred by 50 caliber machine guns, 500 lb. bomb and artillery shrapnel seen in Normandy. However, in 1944 we did business along tree lined lanes like this. The old Jug was chief instrument in avoiding the same soured stalemate of 1914-1918.

Experiencing strafing attacks in Normandy by German aircraft and twice as a POW by American fighters heightened my agility and gets my vote for scare factor supreme. Today, seeing it from the ground gives me a clearer feeling of what it must have been like for both, those we supported and the enemy we shattered.

On the lighter side, I remember the elation I enjoyed on my flitting flight in the L-4 Cub in 1944, same route.

I am thrilled with my send off and the Richard Halliburton slant to the flight before me. I circle up from the Renault car factory parking lot in Paris. It is more than being on my way to Reims. (Our same destination today in 1984.)

Alan Campbell is a talented writer and knows and works with world class contemporaries in New York, Hollywood, London, and Paris. Several of his Hollywood friends, now in the service, are filming and writing in Paris and of its liberation. Alan and friends come to see me off. I probably have one of the greatest send-offs ever given a Second Lieutenant in an L-4 Cub from a Parisian parking lot! A full-bird Colonel props my 65 hp Continental engine. I orbit up and head east. Aisne-Marne canal, Chateau Thierry; all things I read about in "Flying Aces." I seldom missed an issue through high school in Arlington, my home town, population 521.

The little Cub is light and spirited without Campbell and his luggage. I am traveling with the minimum; orders, camera, tooth brush, socks and underwear in a musette bag. I rubberneck; climbing, I go higher and higher, then come to reality; slip back down just above the trees. I sure don't want to have the squadron's only two-seater shot up by some ME-109 pilot looking for an easy mark. Higher and higher, then down again, I soar and swoop.

It seems almost a fantasy. From the Cub checkout with Bill Fisher on, the trip is a lark. The little plane climbs again, I gander, circle the Cathedral at Reims and follow the highway about twenty kilometers toward Verdun to American airstrip A-79 near Prosnes.

After delivering Campbell in Paris, the other reason for the trip is my designated duty to be at the runway when our planes come in a couple of days later. In the meantime, I take advanced echelon squadron people for rides in the L-4 and explore, on foot, the World War I battleground surrounding the runway. The old abandoned land is pocked with shell holes and soil slumping trenches. Rusted pieces of helmets, unexploded shells, dented canteens, and churned and caked earth extend acre after acre for mile after mile. Quiet desolation. A few scrub bushes and spears of motley grass soften the shell holes.

I am alongside the runway, expecting the planes momentarily, when a big old French rabbit lopes across the wire and tarpaper strip going right to left, heading for "no man's land." This is a real challenge. First, are there airplanes in the air? Second, will I shoot a red flare or a green flare? After lightning calculations, I fire a green flare, fairly low trajectory meant to lengthen his stride and hurry him back to the past. I have used many methods in bagging countless rabbits. This is a first for me, to only expedite a rabbit's travel on purpose.

Now, here I am forty years later, with Peg and a group of friends, rolling along the highway to Reims, almost another dream. Everywhere, we see monuments of World War I. It must have been a terrible war; all wars are terrible. Must wars happen? In early school years I would lie on my stomach on the floor, looking at our heavy 1918 War Book. We also had some World War I mementoes; Uncle Mark's leggings and a couple of empty artillery shells.

Today, we drive out of Reims to the little town of Prosnes, near our airstrip site. At a school yard, we are directed to the Chateau Sept-Saulx.

The officers of the 362nd moved from tents to the Chateau shortly after we arrived at A-79. We are here when Chris is shot down and lost. Our last time funning together is a flight in the little Cub around the immediate area. We also lost Conatser and Ort. It is a doggone serious time. I thought Chris was safe somewhere. I wasn't on the mission. Flack! Chris's ship continued, dropped his bombs and went into the ground. His mates figured he tried to belly land under the clouds.

I contacted his family after the war. His Dad and Mother are not much different from my family. We visited them in Modesto, the fall of 1945.

The foxhole digging foursome of Normandy had become productive veterans in the outfit. Losing Chris is tough. War becomes complicated. It is more difficult to be close to newcomers.

We continued to fly missions of ground support at the front and railway disruption into Germany where ground fire became more intensive as "Jerry" pulled back. We gave their gunners experience.

We are element leaders and occasionally led flights. We also led indoctrination formations for new pilots coming into the squadron. It is a time of responsibility, teaching, and learning. In four weeks, one would be leading squadron missions, a second would be doing a stint at ground control on front lines, and I, a POW in a German hospital.

One of my experiences as flight leader is an underwhelming success. I am leading Green flight, the designation for the four ships on runway alert. We are getting infiltrating enemy aircraft on occasion. Runway alert is one of those necessary jobs. Older aircraft are assigned; usually razor backs with lots of hours. We park to the side of the end of the runway, at the ready with warm engines. We are to scramble if "bogies" (unidentified aircraft) enter our zone. The cue is by radio or a warning flare, hopefully before coming under attack by bandits, code name for enemy fighters.

It isn't a bad way to give fledgling flight leaders a chance to think about responsibility and leadership. My "command" is doing its first run up; we start engines every thirty minutes in the October air, when Green four's spark plugs foul, and he taxies back to the line; an hour and 30 minutes to go.

On second run up, another Jug's cylinders fail to clean out; now we are two. On the third run up, the third member of our flight smokes and coughs back to the line. Mine, Green leader's T-bolt, revs and purrs with all eighteen, powerful and clean.

I feel mildly lonesome. My pride dwindles with each member's departure to the flight line. I am extra alert and alone in the lowering overcast. The visibility is O K. Only one ship is ready, mine.

I have been in like position before as a high school freshman. Doc Severson is our band's nine year-old first trumpet and not much taller than his horn. I play the base drum. We are backing Doc in his no-holds-barred triple-tongue challenge to the best in State. I am nervous and "little Doc" is charging like the champ he is. I had taken piano lessons via riding five miles horseback. I forgot much by the time I got back home. Nevertheless, they needed someone who knew a little. I qualified a lot. They settled for a conscripted volunteer, me. I started playing in the band and started, first team, in my first football game, a freshman, September, 1937.

The director, Clyde Simpson, doesn't tell me until the band is on stage in the University of Oregon's huge MacArthur court, "The base drum roll is out. You will do it on tympani provided on stage."

I ask, "What is a tympani?"

The director says, "No sweat. You can do it!"

I didn't want to let little Doc down. I did it, and Doc is the winner. I had watched the other bands before me. My developing philosophy is helpful, has been ever since. "Can't be too tough, look who's doing it."

Pay attention.

I am not surprised when "Gayname," 362nd tower's code name, gives Green flight the order to scramble. I roar solo down the runway. The "tower" gives Green leader direction. I respond in my best four-ship flight leader's voice. "Green leader. Roger, 'Gayname'."

The old razor back leaps into the air, no bombs, no extra tanks, no rockets, just guns; no join up, just go. I search and scrape clouds for altitude, an advantage. I didn't find a bogie. Probably some limping home, late mission survivor.

It is the only time I ever fly my ship. I am proud of her, E4-U, and her ground crew. I am glad to have worked up to having a ship of my own. It has a cowl painting, a "Rarey Bird" wearing boxing gloves ready to jab or throw a right, with the name "Spunky" underneath. I planned on "Sunny" for my ship. Both names fit my Peggy. We left the ship "Spunky."

"Dad" Rarey's story is of delight and sadness. He was a legend to us because of his art. He was lost after D Day, a month before we arrived. I learned he was an original pilot in the 379th squadron, and an extraordinary man; commercial artist, loving husband and father; a tragic loss to Group mates. He combined his creative skills with words and brushes doing cowl art on many of the Group's planes. His family and compatriots joined in publishing Captain George Rarey's, *Laughter and Tears*, in 1996: A combat pilot's sketch book, letters home and with his special philosophy and humor.

Today's memories are a mix of joy and sorrow. It is good to see where our airstrip had been. The World War I battlefields, the trenches, rusty helmets, and rusty canteens are gone since 1944; wheat is growing, and so is a good-looking crop of barley. A new memorial, but dedicated to World War I, is near where our runway crossed a side-road. We hold a memorial service. I had lost close friends here.

With help from school kids, we find the Chateau Sept Saulx. Charles Mignot, the grandson of the founder, walks from the Chateau, greeting us. We are pleased. Fern and Chuck had written, telling the day we would stop by, however we departed from home before receiving their response. Our anticipation is heightened as a French Mirage fighter flies low, level, and fast over the gate behind us as we walk the fifty yards to the front steps. The timing is perfect. Clout? Coincidence? Unexplained.

Charles Mignot extends a warm welcome with his invitation to stroll over the grounds and to come inside. I lap it up. We feel privileged and indebted to our host. He has arrived from his Paris home this morning, especially to open their Chateau for our pleasure.

We appreciate his generosity and his thoughtful presence. He is twenty-eight years old and looks so young. In 1944, our average age is more like twenty-one. Our "old man," Colonel Joe, was 27 when we were here.

Ed and Genevieve's five year old grandson, Eric, is star player of the proceedings in the Chateau. While we are being served Heidseck Menopole champagne in the library, Eric and his grandfather carry on a conversation. Big Mac says to little Eric, "How do you like this beautiful home?" and "Are you enjoying our trip?" then, "I'll sit here and you sit there."

It is a type of diversion for the protection of the beautiful objects de art and fragile decorations in the room and especially to ease the apprehension of Eric's grandmother as well as all mothers present. Eric is an angelic cherub not unlike his grandfather forty years before. He sits up straight in the huge chair and converses across the room with his grandfather.

In the hallway, I ask one of the serving ladies for the corks from the champagne bottles. She smiles and retrieves them. I wonder if this woman could be the young girl who with her mother and older sister had taken such good care of the Chateau when we were here before. She, if it is she, had hesitated and looked to her mother for approval. Is it alright to accept the bars of soap from the American? They were walking from the Chateau that afternoon in 1944. I was just coming through the gate carrying a paper sack with weekly rations of tobacco, candy, and toiletries. Her mother looked me over pretty well. Never looked at the soap. I smiled and was relieved when she nodded affirmatively. The little girl was probably ten or eleven in November 1944. Now, it sure could be she. Maybe the gift was a gesture in hopes that Chris was alive. It was so typical of Chris to be kind to the French folks.

Grandson Charles speaks English with an American accent. I admire his ease of communication. He invites the four former pilots on a tour to the upper floors. We look down on the formal garden, left to the gatehouse and wall, and finally around to the right over the sun terrace and across the expanse of green lawn to a tree shaded stream. It is elegant and pleasing. We admire paintings and sculptures. One area holds a marble bust of Charles's father and a portrait of his grandfather.

The Chateau was a premier living quarters. We moved in from tents which were set up by German Prisoners of War. All three squadrons and Group Headquarter's personnel lived in the Chateau or its service quarters. We were spread throughout rooms and hallways with one squadron on each floor. The large dining room was our austere mess hall; long tables, tin plates and utensils, and a walk through chow line. The room was also our rec room, theatre, and used to screen combat film.

Peggy and I walk around the grounds, down to the creek. I miss a statue, one that reminded me of another at the street entrance of the Men's Gymnasium at Oregon State in 1941. I ask Charles. He says, "A tree fell on it."

I remember John Hill gave me his First Lieutenant's bars here in the Chateau. I was flying from here when I shot down an ME-109. Some of our best missions were from here. We were in the thick of it. We had visions of the war being over soon.

Our group was a part of the XIXth Tactical Air Command led by General Weyland. Each of its Groups has since been proud to imply full coverage for General Patton's Third Army as it raced across France to Germany. In reality, we did it together in competitive chivalry. We protected, destroyed, and shared both the agony and the joys of this duty. Patton's supply ran out at the outskirts of Metz, so it looked like we were going to be mudding it out over the winter. I am told mud and tents were it. (I wintered across the Rhine.) Group was in process of moving when I failed to return from a mission near the Rhine.

I wrote almost daily to Peggy. Here are excerpts from the last two letters.

"Nov 14 1944 – 'tis another cloudy day without much doing. I'm on runway alert. I'm sitting here in my T-bolt, 'Spunky,' and thinking about you and us and misspelling lots of words. My! I sure do love you! I look out across some of the ground that people fought for in 1918 and wonder if it will happen again in another few years. Certainly it won't happen. This is what they said before you and I are born--We don't want it to happen again. I'm sure those boys on up there with Patton will certainly want to make this the last one of these affairs. Those boys are going through one hell of a mess. It's muddy over here, when you hear mud mentioned, it is meant in the worst way one can imagine. Our boys are going to go through it. I'm sure we've got the stuff, men, equipment, and ability to win the war. It's going to take public interest and the courage to say what one believes, in years to come. There can never be an overlooking of sneaking power again. I reckon that's about all the 'soap boxing' I'll do today. Everything is going swell. I'm in good shape and as ever am hot for yo' body little darling. Being married to you is the most wonderful thing. A good way to tell you how much I love you is to say 'Sunny' I love you every bit as much as you love me. Forever your *devoted husband, David."*

A day later, a much more personal letter ended, *"I reckon my love for you is just the greatest thing in my life (signed) let's cuddle, lots of love, Davey."*

I am shot down; more like shot up, two shells in the belly of the plane as I rounded a break in a wooded canyon at tree top level, by ground fire on November 17th, 1944.

I barely got acquainted with my crew and never got to fly my ship on a mission. I am surprised to learn from my 201 file, the attempted runway alert intercept is omitted from the record, isn't included not even as a take-off and landing.

In order for a crew to get a new ship, some pilot had to get the old one clobbered. The newer planes with bubble canopies and shining silver aluminum are the ones for most combat missions. The good old olive drab ships with the greenhouse canopies blending back into a hard ridge down the back, are fondly called "razor backs." We used them as spare. They ran and purred just fine, just didn't have the visibility and some features of the newer ships. Every one of those old ships is a survivor and the top of the line when it joined the outfit. Come to think of it, Razor Back is a pretty good name. Either rooting hog or flying Jug, both are mostly front end and bristled with destructive power and could evoke both fond and fierce response. Dad took the time to show me some semi-wild Blue Ridge hogs in North Carolina while visiting his home country. I'll never forget my six year old feet plowing through the leaves up to my knees, or the sound and glimpse of squirrels high in the bare trees, or the hogs competing for fallen nuts. Dad took time to show me the land.

The next year, he took me to Portland, Oregon to show me airplanes. I am seven for our first and only flight together.

Chapter 34:
Somber

A half mile down the road towards Verdun, we pass a stark tall cross, one I photographed in World War II. There are other war memorials along this road. One is a little fenced cemetery on the right. George says it is German. The grave markers are crosses, small, narrow, and black.

We lunch in the village of Cleremont-en-Argonne; in a main street Cafe and restaurant combination, two steps up from the sidewalk. Our coach is parked along the curb. Our "elite diners" are "in high grass" in the restaurant section. Peg and I eat *jambon* and *fromag*e sandwiches, ham and cheese, in the cafe up front. Through lace curtains, we enjoy the main street outside and several non-tourists inside. I set the camera; outwardly taking our self portrait. I am really trying to snap the picturesque, "lilac time" type, French gentleman sitting at the table behind us. With studied nonchalance, I miscalculate the direction and shutter's timer. The photo's focus is of my mug, me taking a large first bite. No doubt who is the character.

We are the first of the cafe contingent to finish lunch. I pay and walk through to the restaurant and chide our provisional gentry. Their fare looks tasty, rabbit ragout. I didn't tell them I had probably encountered their entree's ancestor a few miles down the road, forty years before. I am glad I arched the green flare passing him on across A-79's runway into the adjoining WW I no-man's land. We volley a couple more quips. It is obvious that we have some time before the leisurely finish dining. I mention, "We'll be strolling on Main Street."

Stepping down to the sidewalk, we skirt around our coach and look across the street/highway, nowadays a back road but on the main thoroughfare for at least six hundred years before. This was the Red Ball highway in 1944. The American Third Army thrived on supplies trucked over this route. Our P-47 fighters supported Patton's armor in beating up the German units retreating along here. The 377th squadron's rocket and bomb packing Thunderbolt's specialty was winning tank duels.

This village is in the heart of World War One's soggy and tortured battlefields. Spads, Nieuports, and Fokkers, forerunners to Thunderbolts, Stukas, and Mosquitoes, battled in the air above, while shells, mortars, tunnels, trenches, mules and men churned the soil below.

A small church sits atop a low hill in front of us. It looks like a miniature cathedral. A stone lined roadway curves up to the sanctuary which has an artillery shell motif. Incongruent? I'm not sure.

Under the hill, the town square lays flat before us. A World War I memorial ages next to the hill. We cross the street, stand in the square, feel respect for those of that other war. My first feelings barely hint the tremor of battle; distant, muted, not personal.

I look to the west end of the square. White marble draws me: A sculptured scene, incredibly powerful; the date is late spring 1944; a starved figure is lifted gently from barbed wire, legs and shoulders, by two others. I don't need interpretation. The meaning sinks in deeper and deeper. I feel, within, a tremendous roaring; my throat tightens, sounds are of fife and drum; words remembered, "Give me liberty or give me death." I think, "And the ability to tell the difference." Freedom is living until death. Loss of freedom is dying until death.

Within, I see my beloved grade school teachers, one playing the piano, the other leading the singing; both rooms, all eight grades gather for singing. We learned about the Pilgrim Fathers, the Revolutionary War also the Civil War; we took turns leading the salute to the flag, all acts and sounds of our freedom. Why do we do these other things to fellow members of the human race?

The inscription is in French. Peg and I work it out together. It is something like:

"100 men and boys were taken from this village and the neighborhood, in retaliation... 72 never returned."

Cattle stampede; blindly follow the leader. Sheep will pile up in a corner and die. A badger will kill fifty chickens for the fun of it; we accept these actions from animals. What time and at what degree can we say "no" to such actions among humans?

The 362nd was just on the other side of the English Channel; was in the skies overhead here, even then. This happened during our watch, only weeks before D-Day.

My uncles were here in 1918. Peg's father was on the way overseas when that war ended. My father, thirty-seven, had volunteered; would go in the next call up. Only two decades between theirs and our war? Seemed more distant in my youth.

Forty years have passed since the happening portrayed before us. Is waging war the only process for waging peace?

Peg and I contemplate going to the mountainous country of my capture and first hospitalization at Kaiserslautern. However, we will do that sojourn another time. We do not want to break the continuity or this wonderful sequence with our comrades.

I will not forget those good people in Kaiserslautern and their caring for me in their packed hospital corridor. Nor will I forget the village of Cleremont-en-Argonne or their caring moving message and their pleasantness to us this day. We walk on up the street to their branch of the bank, *Credit Agricole.* We relate with this farming community and to their land.

I half nap, in the luxury of the moving coach, while George explains the finer points of local history and the things we will see at the French cemetery, *Douamont* of World War I, next before we go to the city of Verdun.

We drive on an upward winding road through forested battleground to the huge monument. Its design is not peaceful; its vertical center column could have been molded in the throat of a volcano; its laterals formed in lava tubes. It stretches wide on a ridge top. The dominant marker, a huge massive central spire, looms looking like a mammoth artillery shell on the skyline.

Detached, at first, Peg and I walk downhill on a broad grassy path. On each side are acres of crosses, each tended by a growing red rose. It is not the same as the American cemetery in Normandy. Here, the names on the crosses all have the same date, in 1916, hundreds of graves. Then another year and day with consecutive squads of crosses. They died in waves and by the thousands. The impact of frightful reality comes more slowly than at Normandy.

Peg is walking along a parallel row of crosses, reading death on another day. This time, I don't hear the roar. I feel the ooze and the mud and the misery and the futility. I think horses--men--cold. It means more when you've experienced even a little mud and struggling in misery.

I look down on hills from Reims to Verdun. When we flew over these World War I battlefields in 1944, they were scarred and barren. Now, they are green with grain and timber. Perhaps there is hope. I expected this journey to have both tender and wrathful moments. I didn't expect this day's powerful exposure. Its stark, not foreseen, reality pounds hard. Surely, others too, have felt the same.

How deeply must we crash before we learn to really be concerned about people and the land? We learned of lands in grade school. We think of climate, geography, and people. Together they make a land. We will not conserve or restore the land until we realize the price of not doing so. Is this basic understanding beginning to show here, near Verdun, in this heavily used country? Must it be terrible everywhere before we see? How terrible must it be?

We walk, individually, for some minutes through the burial grounds, eventually drifting toward the monument's massive spire. Peg sees tears in Fern's eyes when they join us near the steps. Chuck and I walk up to the entrance together, talking, our hearts are loaded and our voices a little constrained. Inside, the attendant indicates quiet. We remove our berets. This is a place for deep remembrance, of contemplation for the future. Surprise continues. The attendant was merely the quiet sign. The speed bumps of learning continue. George told us the setting and the message of this afternoon would be profound. His modulation and serious eyes hinted the impact. He had tried, but we still are unprepared.

We find the striking *Douamont* monument to be an ossuary. Heroes are honored with their bones interred in individual crypts. We also view windowed vaults containing mounds of chalk white unidentified human bones. The horse bones have been separated out. I am a little numb and soon join most of the others on the coach. A few are still in the souvenir and display room getting items to remember. I walk quickly through the room believing, "I will always remember. No need to be reminded."

Back on the coach, someone tells of 3-D battlefield pictures in viewers. I am interested in the recovery process showing here on this ravaged land. Perhaps I can learn from these old photos. Quickly, I go down to the display area. I will buy a couple of pictures. Friends on the bus encourage me to hurry friends looking, so we will have more time in

Verdun.

The 3 D viewers are mostly in use by our missing, now somber, contingent. The scenes of World War I pull and hold. Terrible, wood and cloth airplanes and anti-aircraft batteries intermingle with trenches and mud and shell holes; wounded and dead and dying; men and horses, body and carcass, all in a moonscape of barren treeless misery and destruction.

Barbed wire, no man's land, flat plains, rolling hills, gullies and slopes; shell holes and shell holes on shell holes; this barren land had been pummeled and pummeled again. There was nothing growing; very little standing, only a few broken tree trunks. It has been seven decades since the cleats and boots of war churned this land. I, like the others, am captured. My hurry is replaced with awe and fury. Each new emissary from the bus comes in, spurring us on with, "Let's get going," and is drawn in like he is magnetized to the pictured story. Now I understand the meaning in George's voice when he told about the cemetery and the memorials. Vaguely, I recall hearing him say, "Trench of Bayonets."

Next, we visit the Trench of Bayonets. A platoon had suffered a direct hit from a very large shell. Their bayonets were still fixed on rusted rifle barrels piercing the air above their mudded tomb. The trench-grave is filled by gold-red clay soil and guarded by rusty barbed wire. A huge concrete monolith memorial spans the scene. A frightful, real dramatization of man's frailty: Possibly only a farmer, engineer, or grave digger can really decipher the sequence of saturated mire. The impact, slippage, and the settling occurred in a few enfolding seconds followed by a spared comrade's futile gaze.

The humus robbed, cemented, and humanized soil delivers many silent reprimands. Can we comprehend their meaning? The drive downhill towards Verdun takes us through canopied trees, thriving, with a thick understory. The shell battered land is quiet save for a solitary man trimming in a city block sized clearing with an engine driven weed eater. We stop and walk through the clipped safe area; no World War I unexploded shells, showing now just shell hole on top of shell hole, a pattern of moistened miniature basins.

I am intrigued with the possibility of using the waffled, shell pocked design for increasing water-holding capacity of other tortured lands. The tragic fertility seemed to enhance the rebirth of this soil.

My pictures will show a system that encourages regrowth, harvest, conservation, and continuity. This land bears a message for a sustained future. Do we first have to take the intermediate steps of total destruction? Flying over the watersheds of home, I look down and wonder.

Hans drove our subdued platoon of pilgrims into Verdun, across the canal, and up to Hotel Bellevue. Attractive oil paintings hang in the hallways. Outside, statues of youthful figures and flowers are pleasant; nonetheless, displays of pansies are dwarfed by huge artillery shells interspersed along garden pathways, reminding us of recent images. Our feelings are clouded; we will walk; give Verdun a look anyhow.

We will never forget the scenes of this day. We'll strive to retain the lessons. We are working at translating the message engraved on this recovering ground. I think I read from the tragedy of mud and bones and shell holes of that earlier time, a possible matrix for a sustained future. Are regrowth, respect, response, love and peace with others found only after finding respect, response, regrowth, love and peace with the land? Can these be attained? Neither the land nor its people are permanent as long as man continues his erosive treatment of each.

Chapter 36:
It's Tough, Lighten Up

Peg and I stand on the porch of the Hotel Bellevue not free from our day's tragic revelations. We want to walk, and wonder which way when a painfully bent-over little lady comes by. Her misery does not slow her aged steps, but she does not give us any cheer; admiration, speculation, and compassion, yes. Today is still a heavy scene.

Next, a miniature cement mixer drives by. It is about one-eighth the size of any transit mix machine I have seen. The driver is chugging along with an open air pleasant mien. I am cheered a little.

We walk down by the canal, view rock walls and flower beds. Barges alongside are tethered together. They could have been targets in '44. The twin towers, said to be models for U.S. Army Corps of Engineers insignia, are an interesting surprise. I wore the insignia as an ROTC engineering student at Oregon State, pre-Pearl Harbor.

We study the Iron Gate's mechanism. It once protected the city. Its chains and works seem intact.

Shaped earthen works indicate former fortifications. We are unwinding a little from the recent lesson in rancorous history. Man's struggles seem to go on and on. Do we make progress? Can we define progress in human relationships? Are both quality of life and equality of life among the world's citizens, even a nation's, a state's, or a family attainable? Who will chance it?

Peg and I find ourselves walking into the business district with shops of all kinds; it looks active. Ordinary people are going about living. We go into a corner store. Started plants are displayed in the entryway; garden tools, packets of seed, and all sorts of planting and garden needs similar to those at home. We feel more at ease. We can understand what we are seeing. Peg buys a gift for her mother.

We walk into a combination gift shop and bookstore. The upper level opens to the main street; lower level exits on the next street down. We browse our way; enjoying the change of pace. Outside, we walk and wave to other 362nd walkers.

A shop that appears to sell draperies and curtains also has unique, hair-on-hide wall hangings shaped as animal silhouettes visible from the street. Maybe one would roll up and carry safely in our luggage.

Inside, Peg and I are alone for a minute or more. We enjoy the quiet and try to discern what we are seeing; both materials and enterprise are involved. Eventually, a young lady comes out to greet us. She walks with a lilt that is most attractive. I imagine she is about twenty and a relative of the shop proprietress who comes to the top of the stairs in an open loft.

Even though one of Peg's great-grandfathers was French, we cannot speak French. All I know is *"pomme de terre, merci, and encore."* I think, "potato, thanks, and more of the same!" I really don't want to get my message across quickly. I am enjoying the sensation transmitted by the young lady's delightful walk as she shuttles back and forth to consult with the proprietress. It makes no difference, forward, backward, or standing still, almost, *mademoiselle* is pleasing. Smile, voice, and charm are gifted to this lovely girl. I am reminded of forty years back to a similar animated discussion over the proper size for an intimate garment for Peg. Memory works, with much pleasure. Three salesgirls at a Paris lingerie counter asking in turn, with exceptionally appropriate hands position, *"Mois, monsieur?"*

I, squinting casually, repeated to each, "Madame is petite."

Finally, I, with a spark of ingenuity, asked for a pencil. One responded with *"Crayon, monsieur?"* I loved the round puckered "O" sound with her modulated nasal "N" attachment. Being a fledgling engineer, I converted inches to centimeters and handed the paper to the young lady with the number size computed for Peggy's black lace when-I-get-home gift. Next day, I picked up the finished morsel and mailed it home.

That youthful and former genius is now forty years older (1984): Now, he is in the presence of the "petite" Madame. We select a neat wall hanging. The item is several feet away, high on the wall, behind a display of stacked decorative pillows, and not too readily accessible. The selection is a beautiful bobcat so I say, "Feline?" Does not register. Then I think of the cat back at the Chateau du Molay Peggy's and my first morning in France. She, the cat, responded to the tone of Peg's voice. So the ever-thinking, graying Jug pilot in his best animation simulates a cat purring, "Prrrr purrrrr purrrrrr."

This athletic charmette lights up and sweetly responds, "Mew! Mew!" Kicks off her shoes, floats up over the pillows, and bounces back with the cat portrait and my immeasurable gratitude.

I want to applaud, not only the successful communication but also for her warmth and appeal. For Peg and me, the day brightens. We have a most delightful time in Verdun. The old Hotel Bellevue is cordial and romantic.

Next morning we drive easily, not much traffic. It is not the same heavily-used Red Ball highway of 1944. It is a quiet secondary road. Crops are planted and growing. Harvest is a few months away.

Patton roared full throttle through here with us, his winged artillery, hand-delivering war's missiles and projectiles ala P-47; tough on the messengers but a real benefit to the men and ground units we championed.

We approach the airfield near Etain. Chuck Mann and Bill Plummer begin showing the symptoms visible previously with Ed MacLean at Headcorn and Wormingford in England, and me at Balleroy, Rennes, and Reims in France.

We four pilots were all at Reims, but in different roles. Mac, an original pilot, was rotated home from Reims. According to Mac, his combat time was completed early. "Spent most of the time in the Chateau."

Mac, Genevieve, and grandson, bright little Eric, leave us at Verdun.

Reims was the maturing ground for me and for squadron responsibility. Bill and Chuck were indoctrinated at Reims but Etain was their predominate experience.

A couple of times I am brought up short by the veterans of Etain. "You didn't have to go through that awful winter at Etain."

I readily agree, "No, I didn't. I managed a couple of hardships of my own on the other side of the Rhine, on the Elbe. However, barbed wire did shade my enjoyment some."

Our visit to Etain is not so intense for me. I am a spectator. The players are Paul, Group information officer; Bob, crew chief; Chuck, pilot; Bill, pilot; Manni, armament; Nat, Crew Chief; Ken, materials officer; Claude, motor pool; Randal, armament; and Walt, Squadron exec. The troops spot the city water tower as we enter Rouvres near Etain. They remember the small memorial, still along the road. I eye runway-mesh fence between two buildings. Rouvres is different than our visit to Omaha Beach. That was heavy. This is light and fun.

Hans stops the bus. Chuck and George engage citizens in conversation. Front windows of homes open. French spectators gather. Paul, Walt, and Bill stand on the coffee shop steps, each wearing a different colored beret, and each pointing in a different direction. All are talking. It seems no one listens. I enjoy a cup of coffee. This is going to take a little time.

Back at the coach, chaos is becoming less rampant, although skullduggery is obviously cooking. Hans parks in the center of town. People are ready. Townspeople, a few military types from the nearby French air base, city officials, citizens and all gather. Everyone seems to be players in pleasantry. The village church is across the street.

It is picture time. The Mayor, Antoine Granthil, his lovely wife, Elizabeth, the Mitchell's son, Bruce, and their daughter, Nona, who almost never are in pictures, a couple of itinerant military Frenchmen, a man in a window, our recently arriving hosts for a tour of the airdrome, George, our genial guide, and our humorous driver, Hans. All are in the picture. I have feelings of content when I see the photo. No one is left out.

The antics turned out to be a less formal, more mirthful, presentation by Mayor Granthil, of several French farming magazines, to Childs, the *agriculteur.*

Following the magazine presentation, we drive to the airfield. We get a drive around tour of the entire base. We are asked to take no pictures. The pilots are given a tour of the control tower. We spend a short time "shooting the breeze". We make no attempt to translate "shooting the breeze". Not much is happening. The day has turned very foggy. However, our reception is clear and gracious. We look out through fog from the glass top of the tower. While talking with our counterparts forty years hence, I feel a tint of comradeship, a time bridge focuses fleetingly. Then it is gone. We can relate to them, but they not to us.

Theirs is a combat-helicopter operation. Missions are similar to ours. I sense combat readiness with purpose. Later in our pilgrimage, I sense the same feeling at the West German base near Straubing. Comparable type young men are at both bases: Bilingual, intelligent, well educated, and aware of what's going on.

Preparedness is not the same in a country that hasn't had its homes and civilian population battered around for a while. I'm not sure if the difference is the result of resignation or determination. Is the difference because of the back-'em-up attitude of the civilians? I don't know if their attitude is to prevent or retaliate. The people of England, France, and Germany yet understand and remember the travail of on-site war.

We locate the approximate end of our World War II runway; the control tower is barely visible through the fog; Paul leads us in a memorial service; again a time for reflection and remembering.

Dear Father, we are at a very special place in our hearts and memories. It's a special place for you too, as all places in your creation are special along with the people who man it. We commend the spirits of our departed comrades to thee; their death was not the end. Thank you again for bringing us so far and so long down the years to this very special place. Each of us remembers faces that are no longer with us, associated with Etain, and words again do not measure up to the emotions within our hearts. Give us the wisdom to know which way you would have us go. We ask thy Blessing. Amen.

It takes a little extra time to recall and name long dormant events. Ones that run the full gamut of living and dying, joy and sorrow, then and now. Here, too, we leave a little soon.

Chapter 37:
Washed Vision

We didn't think about the cost of war in 1944, but we knew freedom is priceless. Our objective is to finish the task and get home and on with our lives.

Young men make the best fighters. Is it old men who make wars? I don't think so. I was young when Dad told me, "Son, don't fight unless you have to. If you decide you have to, don't fool around; make your first lick a good one."

Growing up, I kept changing my interpretation of his words. Key words are not the last part. The first part, "Don't fight unless you have to," implies you possess values, diplomacy, charity, and restraint, and are capable of using each.

His "If you decide you have to..." implies you are prepared morally and physically, either to back up the first part, or to carry out the last. Sound simple? No. Only if one skips detail.

My father lived by his creed. He had the right personal makeup to handle both the first and the last part. This is the meaning of "right stuff." Can a nation live by this creed? Possibly, this seems to be the unspoken aim of our people, though many times it is the extremists of both the first part and the last part, who speak the loudest.

Our nation has not physically fought on our own soil with outsiders for at least ten generations. I cannot help comparing the fighting men and women at the different military bases visited in Europe; French at Etain, American at Frankfurt, and German at Straubing (later).

At the Rhein-Main US Air Base, we didn't come in contact with the counterparts of those men we talked with at the French base. Here, we were guided by a you-would-know-him-anywhere public relations officer, cordial and knowledgeable. The base commander seems ready and capable. He registers knowledge of both the local and world-wide

Washed Vision

situation.

The French helicopter unit has a mission similar to ours forty years before, only in a stand-by mode. We were fighting and liberating. We were younger but were seeing things at the same gut level.

In contrasting the military bases, French, American, and West German, I get the impression the French and German bases are parochial, while the American is global. Have we truly made the switch to global vision? Is freedom of thought, from hunger, from servitude, from imposed government, freedoms for which we are committed to? Do we have the capability, stamina, and moral courage to know where and if we can make these guarantees?

I saw government by the people work effectively in the POW camps. The older established camp had excellent internal government. The indoctrination into the code of the Kriege (slang for POW in Germany) and the objectives of the Kriege were evident. Those moving away from accepted behavior were few. It is tough to live by oneself. It is very tough to survive by oneself. Mental and physical anguish are better shared.

POW camp was a great evener of men. Those with superior capabilities and training receive reward from accomplishment. Everyone was expected to contribute to his capability.

The Kriege government was of comparatively short duration. By the nature of military selection, men possessed many talents; most were British, Australian, New Zealand, Canadian, and American air crews. The POW experience in European World War II was probably more humane than any conflict since or in other theatres of war, then. I think I learned much about myself and others merely by the way we responded in a variety of situations.

Do we have to be in a crisis? Was Dad saying, know the difference between being an aggressor or a defender without rationalizing your desire?

The Magna Carta, Declaration of Independence, and Constitution; all point towards equality for the individual's rights. We each have different capabilities, but our degree of conscience, honesty, and morality is not restricted by talent; integrity is personal.

Defenders sometimes become users; wealth and prestige are gained or lost with manipulation of others' rights. Our counter to this problem at home has been the ballot box. It works as long as we are informed and participate with reason and integrity.

Ordinary education, inheritance, good looks, size, and stamina should be rewarded by the way we use these gifts, rather than how we stack up with others.

If we learned anything in the testing, training, competitive atmosphere of Army Air Force Cadets, it was that we all have different makeup. Our strengths are different; our surroundings before joining were different. Our genes differ, yet, we have more similarities than might be expected. In combat, we were, collectively, a group of caring, dedicated individuals.

Luxembourg and the American Cemetery are next. We sustained heavy casualties in the area served by Luxembourg Memorial Cemetery. City names on highway signposts still have encounter meaning for aging tactical pilots. Cities are measured by quantities of flak, terrain, type of target, weather, destruction, narrow escapes, and casualties. Metz, Trier, Aachen, Frankfurt, Kaiserslautern, Saarbrucken. Every name prompts recall of nearly forgotten fighting flight.

Much was demanded of the "P-47-Sherman tank" team. We were something like a combat UPS. We guaranteed delivery and on time. One of our pilots was usually on the ground close to the delivery point with "his" armored unit, talking to "his" air unit, asking for and directing close work for his armor and troops.

The P-47 became symbol of both destruction and protection. Graying and balding ground troops remember and speak of the P-47 as though it were a performing protector, a means to get out of predicaments. Many hard hitting ground units roared across France, liberating miles of territory using the Thunderbolt as its "Ace in the Hole."

In retrospect, the old Jug had automatic capabilities; not many, but some forty years ahead of its time. Wasn't it heat seeking? It often responded with instant update, selecting targets, or immediate identification and recall from a friendly situation. On occasion, it could be reasoned that the Jug or a flight of Jugs had something on board appearing to have thought capabilities. More times than not, the Jug could even determine valid hit-targets from valid don't-hit targets. For those who remember its capabilities and humanized qualities, it's hard to imagine such a sophisticated guidance system being on board in our ancient time.

Not since gun powder first propelled projectiles against saber and lance, has there been so dramatic a change in land warfare. The technique was developed by those wearing insignia of both the Allies and Axis. It was democratic freedom that allowed the responding Allies to reciprocate with superior design and production, training and tactics, and the magnificent adaptability of the men in the machines.

Casualties were endured, and those who survived are grateful. In order to survive, not much thought could be spent in mourning for those downed. We who showed up again appreciate the dual load carried by the survivors. It is not easy to pack the belongings of a friend to ship them to his family--even more difficult if you know the folks at home.

Red, I remember. I'll always be grateful that the message to Peggy said "missing." The message, but for you, probably would have said KIA.

Recently, I checked old letters. Red wrote in October 1945. I was at Dibble Hospital in Palo Alto. He confirmed my ship being E4-U and Fisher's as E4-E, which became Chris's Lady Ethel. He also said Archie Billings inherited my plane and that he was shot down in the Battle of the Bulge.

I looked forward to the 1989 reunion in Hammond, Virginia because I knew a squadron mate, Wiz Wisdom, was coming. We had not had contact since November, 1944. Peg and I were on the hotel mezzanine at the top step of the stairs. Seeing Wiz coming up, in company with his wife Ginny, I step slightly aside and wait with grinning anticipation, Peggy stands with me. Wiz, six feet four and maybe more, came level with my name tag before we made eye contact. He says aloud, "I read David Childs."

He looks up. "But he's, uh, you're dead."

The moment brought home reality. His reading my name tag reaction vividly confirms my wartime deduction. The squadron would conclude, "Dave was killed when the plane exploded on impact into a wooded hillside."

Forty years later, Luxembourg Cemetery weighs heavy on survivors. We read the names recorded. They are eternally young in our minds.

Why them? Many casualties happen in overtime of war. The Battle of the Bulge, winter, and spring 1944 and 1945 were almost after the whistle, but the casualties mounted. The list at Luxembourg would be longer if many remains had not been returned to the United States.

General Patton is buried here. He knew the skills and the determination of his men. He also knew his Thunderbolts and Shermans. Experienced was hardly the word for our troops in this war. From Normandy to Etain was barely five months. Patton was aware of the lives his battle combination was saving; he was also aware of the price many were paying. He was proud of the adaptive coordination on the ground and in the air. A fresh wreath lay on General Patton's grave.

I try to read and to record the words of a memorial and can't; Chuck tries and can't. We take pictures. The flag is hanging straight down, wet with mist. Perhaps the Luxembourg cemetery's message for me is the war is concluded and because of those lost, many survive. The 362nd lost ten pilots and one S-2 officer during training; ninety pilots and three enlisted men KIA; several pilots and ground personnel were wounded; Thirty-four pilots became POWs and several escaped or evaded.

Whatever the reason, for me the anguish is deep; the tears in my eyes and the lump in my throat are different than at previous cemeteries. I am grateful, angered, even quieted. The cross currents, four decades later, are mixed with the realities of the intervening years. Now, our children are parents, we, grandparents, some are great grandparents.

Archie E. Billings. His name jumps at me from the page in the cemetery list. His name represents friends who are gone, not only those buried at the American Cemetery in Luxembourg. Archie was such a nice person, with integrity. Wiz and I went on leave together with him. He, like the others, was to be trusted; we laughed and we flew.

We leave in the mist and the rain for Germany. We remove the 362nd Fighter Group banner from the front of our coach. We had been a proud group with brightly painted red tips and spinners on our ships when last here. That time is a memory. We are still a proud and quality group.

Trier is a landmark, also an air mark too--heavy black flack had marked its sky, even above cloud cover.

Hotel Monopol is a new experience, our first in Germany. I didn't expect it to be much different, but as France is different from England, and England differs from home, Memphis is different from Portland.

Germany is also different. The bedding is a down comforter. Not bad, just different.

We walk through the streets; have a lunch-snack and coffee, buy a bottle of German wine. We ask at the desk for glasses and a corkscrew. The cheerful girl attendants quickly produce a tray, glasses, and tablecloth. They seem pleased to help us. I enjoy their pleasant attitudes as much as any on the tour. Only in the large cities did difference show. Even there, people respond to pleasantness with pleasantness.

Our return to Europe is exceeding our most extravagant hopes. My quest is nearing its completion. I have one more mission and this one is very personal and will not be on the agenda for almost another two days. I am saving Kaiserslautern, where I was first hospitalized, and other significant places for another time or maybe not at all. I am beginning to ask George about Furth and Nurnberg.

Chapter 38:
Joy & Maudlin

Germany seems so energetic. The highway is busy. We travel along the Moselle River, passing villages and factories, through woods and over deep gorges. We look down on rooftops and railroad tracks. I find myself calculating approaches and escape routes. Learned and engrained destruction and preservation habits run deep.

Sixteen of us take a Rhine river cruise from Boppard to Bingen. George chooses not to go. However, he mentions quite casually, "It is customary to get a bit 'swizzled' on the Rhine cruise."

We "fraternize" with a group of about thirty area teachers. They board carrying hobo sticks over their shoulders with red bandanas tied to one end holding lunch. We enjoy their singing and fun. Several speak English; some have visited in the United States.

Commerce and movement seem dynamic. Fast trains ply both ways on each side of the Rhine. Picturesque gingerbread house villages are around every bend. We meet ferryboats, barge traffic, and a hydrofoil.

The heavy current reminds me of the old days on the Columbia River. A few miles downstream from Arlington at the little town of Rufus, Combat Engineers trained for crossing the Rhine. Our Columbia River gave realistic training location for the formidable anticipated charge.

The castles and the Lorelei are widely known. We expected them. I was still impressed by the many castles in that short distance. However, the vineyards on the extremely steep hillsides are the dominant feature. Mile after mile of walls hold the soil on the slopes.

I would like to understand the technique and success of the centuries of land use; same with timberland and watersheds throughout Europe and England. I wonder what deterioration rates are now. Have they stabilized watersheds and the soil? Are they truly sustaining the resources and productivity?

Here, timber is very closely managed. My few minutes of evasion in 1944 occurred in a trimmed conifer forest. The streams are highly controlled.

We enjoyed cooked red cabbage with sweet sour sauce and sliced fried potatoes for part of the cruise lunch menu. Our "swizzle scheme" requires two empty wine bottles. We supply one, and obtain a second by brilliant negotiation. We work on our disembarkation script, especially focusing for the benefit of the coach traveling greeters. It seems Paul's *frau und tochter*, as well as George, and the others going with the coach should be properly greeted, in fact, totally impressed. Our chaplain and chief, Paul, wearer of the red beret, is easily the tallest and most conspicuous. We choose him and yours truly to fly the first element. Please excuse the Army Air Corps and nautical mix of talk.

We are to cruise down the gangplank, front pair each carrying an empty fueling container. This may be construed as a high pass. Leftenants Mann and Plummer make up the second element and come down the plank lower; you can expect so.

The rest of the squadron looks fairly sharp. However, the two harpies, 'err sharpies, Peg and Fern, disembark first for a reconnaissance and photographic mission. Their secondary target is George's reaction to the returning squadron. The primary target is pictures of the arrival of the errant.

A pending incidental casualty might be photos of the open mouthed disbelief displayed on faces of daughter and Mom. The errant flight's approach is probably really convincing beyond any shadow of doubt. Our tall chaplain is sinning; I mean singing, off key. They have every right to be mortified.

But the peel-off is a dilly of a "bombshell." (A four ship type of landing pattern.) Instead of vapor trails it was smiles to the tune and words of an Air Force radar song "Show me the way to go home."

The squadron and recovering observers regroup and enjoy a pleasant walk through Bingen, our port of arrival. This town's name also has meaning on the Columbia River. Bingen is on the north shore in the Columbia River Gorge in the State of Washington, has a similar setting to the Rhine Bingen; uses the Deutsch city's motif in its business district.

Peg and I walk close by an old castle. Wild roses are in bloom along the hill. They look like ours in Oregon.

On arriving at the Hoechster Hof Hotel in Frankfurt, plans are developing for dinner at the Frankfurter Hof Hotel downtown. The dinner party included the survivors of the afternoon mission.

The objective is quite simple; get on the subway and ride downtown; get off, walk two blocks; have a wonderful dinner. The return is straight forward also; walk back to subway, get on; ride to station; walk to our hotel; and the mission will be completed. Piece of cake? No. Confusing? Yes. Did we accomplish the objective? Yes. Expeditiously? Up to getting on the subway to home.

This was only after complete confusion, partly enhanced by a North Carolina porker physician (once a Jug pilot) leading us to chaos with his conflicting interpretations of route maps. Everyone is sober. After we try various choices and unravel the problem, we are both cold and sober.

Our problem is when we get things figured out and are ready to board, the train comes roaring in from the wrong direction on the wrong track. Next time the train arrives, we are undecided with our party divided, on both sides of the track. (Only a few feet apart, but almost two blocks away by overpass.) For the people on the right side of the track, the train was going the wrong way. For those on the wrong side of the track, everything was wrong.

Next morning we are a few minutes early, waiting at the entrance of Rhein-Main Air Base, for an arranged tour followed with a briefing by the base Commander. Huge commercial jets are touching down on the other side of the fence less than one hundred yards away. Fascinated, I walk beyond the curbing to watch their approaches and touchdowns. I snap pictures.

I grew up watching airliners and mail planes in the thirties fly over our ranch. Beacon lights blinked into my bedroom window, dimly lighting the Roman pillar trim on the wallpaper. We lived on the south boundary of the transcontinental air route on the leg between Portland and Pasco, WA. One beacon flashes two miles to the north, in a fenced dry pasture with white alkali rocks in the gate cribs. Another is ten miles west in a wheat field in Sherman County, a third, ten miles east perched on a basalt bluff overlooking Willow Creek at the west edge of the

Boardman Desert. Ten miles to the north, Arlington airport's beacon swept white and blinked green.

A neighbor's oldest son, George Newell, started his career caring for local beacon lights and navigation signals. He contributed in air safety during World War II and afterwards, retiring from NASA's space program. His work extended from the Berlin Airlift to include Eagle's liftoff from the moon. His wife, my cousin, believed they lived for a time, in the German village where I was first taken after my capture. The location, timing, and events described make it possible.

Now, marveling at these tremendous planes coming in close and low, throttled back, elevators bigger than wings of the past, thrust reversing, with mammoth capacity for speed and freight, rekindled the awe feeling of my first aviation experience. Am I daydreaming?

I was seven in 1930; the place, Swan Island, Portland's airport in the Willamette River, one hundred-fifty miles west of our ranch. I was cautiously hopeful as Dad asked the man at the counter if we could get a ride in an airplane. I stood looking up against Dad barely seeing the man. He said, "Yes, there is a student instrument flight just going up. They'll take you."

We followed as he hurried through the screen door. I let it bang. He flagged down the instructor and student as they were taxing by.

It was a four-place cabin, high wing monoplane, with bench seat in the rear. (Decades later, I learned it was a Stinson Detroiter.) In the air, Dad pointed down to the Willamette and Columbia Rivers and Mount Tabor near my Aunt Delia's home. I had a queasy stomach for a few moments. Dad saw my ears turning white and got me to looking around at objects in the plane instead of the ground. I sure was glad I didn't have to barf.

A Ford Tri-motor was loading as we walked back across the turf. I looked up to the high wing. The corrugated sides reminded me of our neighbor's barn roof. Seemed like the tire I was standing by was as high as my head. Coming from dry and summer brown Eastern Oregon, the expanse of close clipped lawn grass everywhere around was also pretty impressive.

My next ride, five years later, was in a Cub with floats. I was showing a 4-H Club steer at the Pacific International Livestock Show. Dad said, "Go ahead," when I asked for permission. He believed in living as well as being alive. The plane flew from a dock just over the dike on the Columbia slough by Portland Stockyards. Flying was a lot more fun that time.

The planes that eventually really grabbed me were the little army pursuits. However, the ones flying over the ranch daily were twin engine Douglas airliners, the forerunners of the DC-3 and C-47, a workhorse of the Army Air Force in World War II and for many years afterward. I whittled and carved models of both; mounted them on the corral gateposts, with propellers turning in the wind. I like watching airplanes. Always have.

We are treated cordially with a tour over the Rhein-Main base. We are told not many ordinary citizens come to look-'em-over. It seems our presence is both welcome and appreciated in 1984. Rhein-Main is involved in most congressional European junkets, as well as a staging area for almost everything happening in Europe and the Middle East. Many events that affect our armed forces and citizens outside the continental U.S. have significance to Rhein-Main Air Base. Stories in the news since, many of them tragic, have turned my thoughts back to the American base of that day. Now, in 2002, we often read about Kaiserslautern's Ramstein. It appears the rolls have been reversed.

We appreciate the time given us by the base commander to brief, and talk with us and answer questions. I feel confidence in his ability to make the "know which" decision, of my Dad's philosophy lesson, in whatever form it is handed to his command.

We are given a tour of a desert camouflaged older cargo ship, and asked to confine photographs to only this plane and to carefully align our pictures to exclude all other planes on the flight line.

My positive World War II connection with the Frankfurt area includes a mission flown from Reims A-79 to the east of Frankfurt. It involved air-to-air combat and the disrupting of railroad tracks.

My passive connection with Frankfurt is also important to me. It was from Frankfurt that the 362nd was flying missions connected with liberating POW camps.

Nurnberg and Munich are on our sojourn's schedule. The 362nd was based near each of these cities. I was also in POW camps near both. Memories are surfacing. The depths of each has not been fully visited in thought. Perhaps because I was seriously restricted in physical action by injury and the uncertainty for the outcome.

My young man's goals were given a slight change in direction.

Chapter 39:
Quest

I am anxious, a little apprehensive. Will our Nurnberg quest be a success? George isn't too encouraging, though he is most helpful and paves the way by finding a taxi driver guide. I want to revisit the scene of my 1945 arrival and trek through Nurnberg. The event is clear in my mind.

Furth: Railroad station, the walk, bombings, the stadium, autobahn, prison gate, the barracks.

We have abandoned the idea of finding the site of my bailout and first hospitalization as being too impromptu this first visit back. We do not want to change the mood of the 362nd's return. We especially do not want to miss any of this trip's select places.

Nurnberg on our own will be enough challenge. Peggy, supportive, knows events of those two days and nights have significantly influenced my life.

I experienced the camaraderie and mettle of a small group of individuals who enhanced my creed. (Do your best.)

Nurnberg: Getting there in 1945; my stay; and the leaving; all are stacked together in private remembrance. Overwhelming, describes my desire. Yet, I seem to have a measure of patience. Why is this place so important to me? The bailout, capture, interrogation, injury, bombing, strafing, seeing again, and walking; all were emotional highs and lows.

Why then, after thirty-nine years and a couple of months, do the miles and events between the Furth passenger station and Nurnberg prisoner-of-war compound, Stalag XIII-D, engross me so? Is it pride, humility, achievement, sorrow, trial, patriotism, brotherly love? The answer is yes.

We arrive around four o'clock in the afternoon, quickly check into our room and return to the desk of the Hotel *Am Sterntor*. George is prompt. We desire a guide who knows the city and knows about World War II conditions then. The person should also be able to talk with us in English. The arrangement is made by telephone. George leaves us on our own. Our preparatory research is nonexistent. Our guidance is really only from memory and hope.

At five o'clock, a slender young man in a white shirt, a taxi driver, comes to the desk asking for Mr. Childs. He has an accent, yet is easy to communicate with. He obviously was born twenty years after I left Nurnberg.

I explain I want to see the Nurnberg Stadium (Zeppelin Field) where the big Nazi rallies were held; I want to start at the Furth railroad station and travel across town to the stadium. Then I want to go on beyond the stadium a half-mile or so to the site of the World War II POW camp. I think, in all probability, the stadium has been demolished and the camp area completely changed. I ask, "Can we possibly find something I will remember?"

The young man listens intently and asks, "Was the stadium by a lake?"

I respond, "I don't know." I think it might be because in February 1945, when we neared the stadium, about daylight, the fog was very thick. I ask if there is any remnant left.

He tells me, "What you want to see is there."

I cannot respond though I hear clearly. He also says nearby is a large football (soccer) stadium and there is a game just about over. If we go now, we will be caught in the traffic. It wouldn't be easy to get where we want to go. Our young guide recommends we wait an hour. He also tells us at six o'clock he has to turn his cab over to another driver, therefore he will be unable to take us on the tour. I must have shown doubt or confusion. He says another driver who knows the area even better than he, who can also speak American, will help you. He assures us the new driver will pick us up at six o'clock, and we can go on our tour. With handshake and a nearly doubtful "thank you," we say goodbye to our only tie to our quest.

Peg and I are just beginning to sort out all of the things talked of and the arrangements, when through the lobby door again walks our young friend. His message is, "Not to worry, the other guide will be here and he is more qualified."

His purpose in returning is to tell us not to expect the other person until about fifteen minutes after six, because it takes time for the drivers to exchange and service the taxis. His attention to detail and conscientious service puts us at ease. With gratitude in mind, we ask him to drive us to the Nurnberg railway station. We have an hour before our other guide will show up. This affable helper indicates the station is only two blocks away. We can easily make it on our own. I reply, "I am not sure and anyway I don't want to get turned around and waste precious time being lost."

We want a chance to hand him a reward. We are thankful to him for his warm presence and for his contribution to our Nurnberg success.

In the railroad station, we look at model engines. We try to read signs, buy a map of the city, some magazines about steam locomotives and German tanks of World War II vintage. A plaque shows some of the first locomotives operating from Nurnberg.

Peg and I walk back toward the hotel and venture over into the old walled city. Its buildings have been restored to the same plans that existed prior to the demolition by bombing raids; materials are mostly stone and brick.

We purchase steins as gifts for folks at home and walk and enjoy the buildings. We are too preoccupied with our desire and hopes to enjoy this time to the fullest. We are back sitting in the hotel lobby when a man almost my age comes in and in near flawless American inquires for Mr. Childs. He is bareheaded with graying hair combed back. He is of good stature, five foot eight to ten inches tall.

We briefly outline, repeating once again, the journey of my past. I got off a train in Furth, forty years ago; had walked through the town to the Nurnberg Stadium and beyond to the POW camp. He isn't sure about the camp, but the rest will be no problem. I also ask that he record the distance from Furth railroad station to the stadium.

The small white cab is quiet and comfortable. The silence is a little awkward. Soon, the driver pulls over and suggests we might want to photograph the Hall of Justice, Spandau Prison, where the Nurnberg War Trials were held. I think, "Where the sentencing of the convicted Nazis occurred." I say, "I know about the War Trials, but they are not what we are interested in today."

I was in Dibble General Hospital in Palo Alto, California, when the German Axis notables were brought to this building in 1945. I get out; snap a picture only because the driver has stopped. Since then, I've read parts of Albert Speyer's books, written while he was in the brown stone building. He mentions, with all of the destruction, it is remarkable the Hall of Justice went undisturbed through the Allied bombings.

A year later, in October 1946, he and the others were still here, their sentences having just been declared. I was finishing military hospitalization. In the hospital, we didn't follow the war trials closely. We were looking toward the future. I vaguely remembered the sentencing. I had new eyelids and eye brows. I was twenty-three. Albert Speyer was forty-one and looking toward twenty years in prison.

We chat a little more freely and arrive at the Furth railroad station. Peg and I walk through the station and outside. The platform and steps down to the street seem familiar. I look up. Furth. Same place. Same sign. New paint? This time, Peg snaps my picture.

It was night, around 8PM, when we started out from here on February 20, 1945. A German Sergeant, a corporal, and a private escorting we seven American officers from Muhlberg and then through the night during a British raid, through Nurnberg to Stalag XIII-D.

Today, our driver is showing us the route I probably had followed. We have come about seven or eight kilometers and are still going. He describes the dreadful bombing of the city, which was concentrated in two time periods, perhaps a couple months apart.

That night, we walked steadily...stopping once; just long enough for the Sergeant to convey he would not be responsible for our protection if we were in the city after daylight. We set the pace.

The morning was February 21, 1945. The same date Peg received my first POW letter.

Soon the traffic is thinning. We learn our guide has lived in Italy and has been around American bases and construction.

There it is; the huge wall; the central gates; a boy and girl are playing racket ball; the swastika is gone. Darn near a quarter-mile long; the doors and the steps.

How come we're not stopping? We drive on down to the far end; see a rough painted peace symbol on the wall, like under a bridge at home. We go around the end and right up the huge driveway where men and equipment once passed before the reviewing stand.

This was a part of Hitler's beginnings; this mammoth stadium, holding mammoth rallies with mammoth flags; goose-stepping men; rolling war equipment; search lights played upward like giant columns; then two, three, and more hundreds of thousands of voices raised in one gigantic chant. "*Sieg Heil, Sieg Heil, Seig Heil.*"

Even today, it bothers me to write the phrase.

We stop and look up at the iron railing. "That is where he stood?" I ask and state in the same sentence.

"Yes," our guide replies.

I question, "Are you sure?"

The reply is quick, fairly low. "Yes, I'm sure. I was here. I wore the black armband. I believed. I was fourteen."

I think of the armed Hitler youth eyeing me in the air raid shelter under the railroad station, while two P-51s strafed topside. I think back to 1937; I start all twelve games, played left tackle on Arlington High's football team; our coach, Mr. Vince Barrett, was the best; we were Small School, eleven man, State Champs. I, too, believed. I was fourteen.

Rain is falling now. I look up once more to the podium and the iron railing high above us. I take pictures, get in the cab and ask the driver if he can take us to the site of the prison camp. He thinks he can. I know it isn't far. I also know the railroad yards were nearby. The RAF Mosquito bombers had bombed them often, even on very dark nights. The spent anti-aircraft shrapnel came down on the camp barracks' roofs.

Just a little way down the road, we pull into a wooded area with a

Quest   287

playground back in the trees. The building fits my description of the commandant's office and the guard's barracks. Our driver waits while I take pictures. He thinks the few buildings remaining are now used for low rent housing. Smiling, he speaks through his rolled down window "They'll think you are from the tax collector's office."

It is raining very hard as we drive back down the autobahn. By the time we drive by the stadium, a rainbow is arching across the eastern sky; the sun is setting. I think the young driver sensed that the older fellow was just what we needed. We are grateful to both of them. When we are almost back to the hotel, the driver says, "It is good you are here. Some of your friends are buried here."

The distance was fifteen kilometers from Furth to the barracks. In the forty-year interval, I've traveled it a few times in memory. Perhaps I won't travel it so often now. I am thoughtful. I don't want to be forward, yet I want a little more meaning with this man. I am pleased when he suggests, "I would like to shake hands."

"So would I."

Our friends are finishing dinner as we come into the dining room. We briefly tell where we have been. We sleep well.

The following morning, much to our surprise, the whole doggone bus drives out to the stadium. We look it over quite closely. I answer "No" when asked to stand for a photo up where Hitler stood. A small tree is growing out of the concrete wall to the left of the stadium entrance. I take its photo. The concrete is crumbling on the reviewing bleachers, and the huge peripheral seating is only on dirt, faced with planks of concrete. The construction was for show. There is some barbwire, and grass and weeds, and kids. The swastikas are gone. I feel silence and a calm.

Chapter 40:
Vinegar and Wine 92

The summons brings me abruptly from deep sleep in featherbed luxury. I grasp for time and place. Remembering spurs my anticipation. Again comes the plaintive call. No mistaking a young calf in distress. A mother cow's answering bawl trumpets across The Valley of the Stour. I roll out and bare-foot across the smooth board floor to the centuries-old window. We had arrived very late in darkness.

Below is a fairy-book scene with mist and dew. I look down from a second story vantage over luxuriant water meadows. Beyond the ancient moat, now dry, I see the cow herd. Their heads set, ears up, necks stretched, they are signaling indignation. The cattle and time of day are familiar. The difference is we are a continent, an ocean, and a few retirement years away from a half-century of responsibility for bristling bovines. Like a committee of mothers converging on city hall, loud and righteous, cows too, let headquarters know when babies need care. The week-old calf had gotten on the wrong side of the fence, couldn't reach breakfast and is now raising a fuss.

This is our first, a barely daylight, glimpse of Garnons and the beginning of two days in Wormingford, an English village near an airstrip from which my World War Two Fighter Group flew its first combat mission. (February 8, 1944).

Peggy and I being the farmers and ranchers of our company of a dozen are paired with farm hosts, Hugh and Gillian Gray. Their eleventh century farm, Garnons, with its magnificent three-story manor house built in 1603 on a knoll with a moat amid surrounding farm lands was, to us, "an undared-for dream come true."

Brits call it corn, Yanks, wheat; their cattle are nearly the same as ours. We, too, grow barley. Sugar is sugar, from either an Eastern Oregon or East Anglia sugar beet field. England and its authors and gardens have pulled at Peggy since her school days.

Ed Maclean's letter in May 1991 hooked us. We had many reasons beckoning us to England. QE2's, two for the price of one was a sure clincher.

"To: Potential volunteers for the inspection and analysis of the QE2 sailing for England May 17, 1992."

Ed Maclean, president of the Ninth Air Force Association was looking for people to appraise Queen Elizabeth Two's suitability to be the hostess to the Ninth's 1994 Reunion in route to the fiftieth anniversary Commemorative events in Normandy on D-Day 1994.

Her forerunner, Queen Elizabeth, delivered the 362nd Fighter Group, sans aircraft, to England by way of Scotland's Firth of Clyde, November 29th 1943. We certainly could count on her luxurious namesake to get us to South Hampton, pampered and raring to go.

We were enthusiastic. Vic Lewis, of Buddies of the Ninth, recently found my Stalag IV-B friend, Jack Lomath. Jack is Cockney; born within sound of the "Bow Bells," East End London. Jack was my mentor and benefactor in December 1944 and January 1945 at Stalag IV-B. We met in the POW camp hospital.

It was wonderful to get Jack's letter. We will see him and his family after we attend memorial services for members of our group lost in combat. We fire back a big "Yes, we will come;" logic says don't miss this one.

My on-board assignment is to check out the medical facilities and the exercise department. The way over was wonderful. Peg and I make the most of classes, dining, lectures and cage golf. The little girl, about nineteen, German, stewardess for our cabin, is quite surprised to find we are aware of her home town. "You have heard of my little village Berchtesgaden?"

We are in England for three purposes. First is to visit Jack Lomath. Our second mission co-shared top billing with the first. Peggy and I will participate in the dedication ceremony for a memorial to the men of the 362nd Fighter Group lost in WWII.

In between achieving the primary goals we want to see some of the wonders and beauty of England and Scotland; especially the land and rivers.

Honoring; needs some knowing. The 362nd Fighter Group, nicknamed Mogin's Maulers, flew from two bases in England, a forward Normandy airstrip, two more strips, and a base in France; finally from two airfields in Germany. The group entered combat in February 1944 and fought through to the end, in May of 1945.

The four of us probably are typical of the second round of replacement pilots; two joined the Air Force directly out of high school and two joined during second year of college. We four joined the 362nd in Normandy. We matured rapidly while crossing France. Red was escort leader taking a glider-borne medical team into encircled Bastogne, finished his tour, lived in Oklahoma, owned and operated an agriculture flight service, also charters for Midwest football's enthusiastic fans.

Don, Kodiak, became "Pilot talking to pilots" with an armored unit for a month, returned to the squadron and finished his tour, stayed in service with air sea rescue, retired after 20 years. Don lives in Northwest Washington State. After retiring, he flew B-17s fighting fire and also in the filming the modern version of the "Memphis Belle;" flew five of these wonderful birds to England, where makeup artists daily turned his hair from white to the black I remember. Now, he and his dog, Loafer, walk two miles daily, then Loafer does his loafing, and Don putters with Model T and Model A Fords.

Chris and I flew in different flights. Unique in our squadron was the fact neither of us were hit by enemy fire. We quit mentioning this pleasure at about our thirtieth mission. Each of us took hard hits his last mission. The mission count was in the forties when he was hit. Chris and I were teamed once to fly as a pair. We each carried two one-thousand pound bombs. Chris led; we were successful.

Chris died in a Thunderbolt's battle crash, the last of October 1944. Our son, Christopher, is named in his honor.

My combat flying was forty-eight missions, Normandy across France into Germany, ending about noon on November 17th 1944. (A tour was between eighty and a hundred missions. Some flew more.)

The 362nd Fighter Group chose Wormingford from all its combat airstrips and fields to be the site honoring comrades; others are Headcorn, England; Balleroy, Normandy; Rennes, Brittany; Prosnes near

Reims; Rouvres, (Etain) beyond Verdun, France; Frankfurt-Rhein-Main, and Straubing, Germany.

The monument, designed with reverence by battle buddies is a brick and mortar column, created in a trapezoid with three fronting faces, and a tilted top. All faces and top are thick stainless steel. The top plaque is beautifully etched, depicting the 362nd flying from the Wormingford airfield; Ninth Air Force insignia; a combat loaded P-47 lifts off, another with ordinance, already delivered, starts 'peel-off' for landing. All eight combat bases are listed.

These inscribed words, bring forth the basis for honor:

EVER REMEMBER THESE AMERICANS WHO GAVE THEIR LIVES FOR ALLIED VICTORY OVER TYRANNY!

Four ascending steps honor the three squadrons, 377th, 378th, 379th and Headquarters. Together, they comprise the 362nd Fighter group, about twelve hundred men.

Three fronting sides of the memorial present stark columns of their names, one hundred nine in all.

One hundred and five pilots, ten of them missing in action, lost in combat. Four ground crewmen also were lost by enemy action. Twenty-nine pilots were prisoners of war. Five pilots and one ground crewmen were lost in the United States during training, in 1943; four were lost in training after the war. Lists begin with the inscription "The ultimate Sacrifice."

I listen in reverence to the words of Minister; then Commanders of Group and Squadron, and the reading of a letter from the Chairman of the Joint Chiefs.

Bugler's taps brought a halo to the names. The notes lifted and fanned the memory embers, unlike I had known before. Afterwards silent moments extended. When a foot shifted, it triggered a firm British voice to speak, because he knew, "Wait for the piper."

The lamenting pipes stir, then smooth; challenge and unite, like nothing else could. But afterwards there is no silence. My eyes become the instrument. The crescendo mounts. I see down the column, reading deeply grooved names, one by one; only at the end do my moistened cheeks sooth and quiet the pain.

We host a *soirée* for our old and new British friends; raise a toast to comrades back home who couldn't be with us and another to our ladies. We adjourn across the road to the airdrome, where we look at a lineup of WWII vehicles restored to a "T." I take pictures of authentic olive-drab trucks and jeeps. One has 362nd Group and 377th Squadron stenciled in white on its bumper.

Two restored L-4 Cubs catch my eye. I had flown the 377th's several times; once used the Eiffel tower as a pylon. I am elated when invited for a flight. If I'd been an old hound, "snoozen and dreamen," I'd look like I am livin' my puppy days again.

This day of vinegar and wine seems to free me from a shackle I never knew I had. We find its joy with friends of then and now.

This morning we gathered at Saint Andrews for worship service. It was celebrating its annual Festival of Flowers with its powerful message, *remembering*, the theme of the decorations; and *connection*, the theme of the sermon. The sanctuary was filled. British and American color guards presented both flags.

The church set the message with skillfully arranged sand bags, helmets, and armbands, reminders of united spirit and the events of those WWII days.

The old, old, flint-stone church with its clock and red tiled roof has seen centuries of joy and sorrow. Are we just a ripple? Would our action of then and thoughts of this day vindicate the price?

Hugh and I toured the farm, saw landscape used in Constable's paintings and swam in their pool before joining Peg and Gillian for evening Songs of Praise service at Saint Andrews. Then "home" to Garnons. There were no problems, no delegation of irate mother cows. It seemed tranquil.

In his letter several weeks before, (we had sent our itinerary) Jack had introduced his and Edith's daughters, Marian and Vivian, and husbands Tony and Jim, also the grandchildren. He had mentioned special plans for us. "Call Marian when you get to London on Monday, Tony's off and has something planned for you." Tony is a manager at Heathrow airport.

This is the time to call to be sure of connections. Marian answers, right off says, "Dave--there is a sad change. Dad is diagnosed terminally ill."

It is a relief to know we will see him, yet sorrow for confirmation of my fear. I hadn't dared to hope too strong; subtle words in his letter were tips that something might be wrong. We set the time to meet Marian in London next day. I have a problem with my swallow, my eyes, too. When we meet, I will do it cheerily but at this moment, I wouldn't bet. New Wormingford friends and old are coming through the arched front door when I hang up the phone. Together, we share a fine evening; Hawaii, Oregon, New York and East Anglia, it's a shrinking world. Peg and I enjoy the food. Each local guest has brought a tasty specialty. It has been a storybook beginning but has become a little tough along the way.

Monday morning we say goodbye at the station. We planned to be with Jack and Edith the following Sunday. Now we plan to move three days sooner to Friday. Marian and Vivian have been taking turns going around the clock helping Mum, look after Jack.

I have brought 1944 pilot's wings as a token gift. Also we've come with mementos to show; Peggy's little Testament, I'd carried it in my shirt pocket; Jack's name is in it. I brought a piece of my shirt tail; not much left after my fiery jump had burned cuffs, collar, and medics slit the sleeves. I had asked Jack to cut off and give me the shirt's tail before he trashed it. I could say, "I got out with my shirt tail."

Peg and I ask each other. How soon? What will be the situation? What can we do?

We settle on design and words; we pin the pilot's wings over a small backing of shirttail with the inscription:

Jack

You once saved a part of my shirt tail & a lot of me. Here are portions of both Cheers-thanks. Thank you, Jack

signed Dave signed Peggy

The phone rings announcing Marian and Tony, in the lobby. At first sight, we feel at ease. We spend an hour talking over tea. I ask Marian to give our love and the token to Jack. She suggests we will see him and Edith on Friday. Jack had written, four weeks before, *"I like it when you say, 'We're fairly spry and adaptable.' Edith & I will have to see if we can get some of that to rub off on us."*

Tony says, "Bring the verse and wings when you come." (I sense he wants to not seem in a hurry.) I don't say, but feel "good-time" is uncertain. I brought the wings for a gift to Jack. I had hoped to share hours with him. Now I only want to support him and family. I hope the tattered offering will let Jack know we care and are coming.

Next to Peggy and family, my pilot's wings mean more to me than anything. They are a part of my life. My fondest picture, of us, is she pinning them on me at Luke Field's graduation. These are not the presentation wings but are of the same vintage. I have not bought new wings since 1945. I swapped the wings I was wearing then with John Evans, an Australian RAF pilot, for his in Nurnberg, Stalag 13-D.

Jack and I saw each other just once before and then for only a few moments. He described the scene in a column he wrote 25 years ago; sent me a copy with his first letter in January 91. I have excerpted it here.

"…That's how I came to meet Dave and Chuck; I've called him Chuck; it seems to be a popular American Christian name. I'm sorry I can't remember his name although I remember Dave quite well …I was used to sights that I have no wish to dwell on, yet when I saw these two chaps in their beds, the lump in my throat seemed bigger than ever before…as I stared at those two figures completely swathed in bandages, so that no part of their head, face, neck or shoulders were visible…These two lads were American fighter pilots…burned…miraculously rescued. If I could think of a stronger expression than suffering I would most certainly use it…I cherished the fact that I had been born a "Cockney" because most of that Tribe as it were are gifted with plenty of spirit, & a good sense of humor and boy! We needed it now!" His story continued with forgivable license and a twist of humor, and concluded with*" …I cannot end this tale without mentioning what could I suppose be referred to as 'The Denouement' or the 'uncovering'."*

... an orderly came to tell us, (Ted and I) that Dave would like us to meet him in the "Op's" room ... Dave was sitting on a trolley, with his legs dangling over the side: the Doc was there also... this was the day that the bandages were to be removed from Dave's eyes, said he wanted us there ... As we entered he held up his hand, and said "Don't either of you say anything." "I don't care where you stand, but one of you stand on this side, & the other one on that." "I want to see if I can recognize which is which with-out you saying a word, alright?"

"Well the unwrapping started you can imagine the tension and trepidation as we waited. Dave waited a minute, then holding up one bandaged hand to shade his peering eyes, did a double turn, looked at us both and said without hesitation, 'You're Jack and you're Ted!' How he did it I'll never know, but relief that all was well was tremendous."

Jack Lomath went to main camp before I had dressings removed other than time in ops.

We picked Viv out easily at Brighton station. Her husband, Jim, is a Major in the military. She teaches. They were based in Germany for several years; now near London.

Jack did more for me than I for him. We ended our earthly acquaintance like we started back in December 1944, with a cup of tea. Only this time it was I shepherding the cups. His voice was strong and eyes cheery--wanted to know if our lunch was done right.

Lunch, with daughter Vivian, at his college was "done right." Lunch was terrific. Several of Jack's former colleagues came by our table, offered tributes, and asked of their friend.

Jack, when in his thirties, (I was twenty-one) fortified my father's philosophy; to the worth of all people. I've carried their lessons for many, many years, and now still new ones. He gave us a book of his poems, with a special hand-written verse to Peggy and me. He read one, strong and proper, from his pillow.

Now, we have traded "big lumps" in the throat. Jack died on Wednesday following our visit Friday. Saw him a minute before, maybe twenty minutes now, forty-eight years later. Minutes or seconds, either are priceless.

We shared a small window in our lives, a few days more than a month and a lifetime of respect and remembering.

We left Brighton with smiles through tears.

We saw much of England for the first time. Southampton to Wormingford to London to York to Edinburgh and Glasgow. London to Bath and Salisbury; London to Brighton to London's underground and a round; Lochs, timber, plains, rivers, cathedrals, forts, palaces and museums--a beautiful land giving many lessons. The fond memories are tumbling still.

Back home the L-4 flight led to flying again, passed the physical and bought a share in a flying club, logged sixty more hours flying and shooting pictures of the land and its water until holes in my sight grounded me two years later. I remembered the swim at Garnons, same day as the L-4 flight. I sold the share in the four-place Piper and put the proceeds into a hot tub in Peg's greenhouse. Dim seeing is great for dual splashing, darn near mandatory in our eighth decade.

Chapter 41:
English Channel
D Day + 50

Queen Elizabeth 2: Statue of Liberty to Southampton. Our crossing is as marvelous as we knew it would be. Walter Cronkite is our Ninth Air Force Association's guest. He boards from our check in balloon arch on the dock in New York. He is first rate, the top notch, totally likeable man we knew him to be. Guest conducting the on board US Merchant Marine Band Is just another of his many talents. Bob Hope and Delores join in Southampton with music of a Glen Miller band. "Thanks for the Memory."

The last three days are special on the water. The QE2 speeds by two of the racing yachts in the World Class Whitbread race. They look classy, lean, long, and lovely.

Andy Rooney, too, is with us, short and cherubic. He wrote wartime good things about the Thunderbolt.

We are pleasantly surprised when yachts in the race pass us going up the west channel approaching Southampton. We are getting into spectacular sailing, it is windy on deck; we seem to be slowing. A racing yacht with billowing sail passes us as though we are standing still.

What gives? We are standing still. No we are not! We are going backward, fast! Locomotives and Caterpillar tractors have Johnson Bars for ahead and back changing of direction. I don't know ships, but I know pull and drag and speed and wind. We get word Cunard's show girl of a lot of years is figuratively hoisting her skirts and reversing back the way we came. The wind is reported too high for her to maneuver on in the west channel. We are backing out and heading around The Isle of Wight, the longer and secure way into port today. Yet, we are docking when two of the race's boats cross the finish line.

The Ninth Air Force Association hosts a reception for our many British friends and Buddies of the Ninth. Next day, Cunard offers several tours. Peg and I choose traveling along the River Avon with its centuries old water meadows, and across Salisbury Plain to the Stonehenge. All reinforce our memory from two years before when we were part of the "guinea pig" crossing just to see if we could hold a Ninth Air Force Association convention on this grand liner. Peg believes the same duck entertained us at Salisbury Cathedral both visits.

Energizing, fabulous and amazing! What a glorious event! Centuries of experience on land and sea, and nearly another in the air seem to merge skill and might of nations, in alliance in celebration of unconditional victory. General George C. Marshall's post war plan insured results, offered aid and constructive diplomacy. We are here for the fiftieth anniversary of D-Day.

Recall blends pride and pomp for us spectators. Most ships participating are still at anchor. Some rare vessels rove within a pack of private craft. Tens, maybe hundreds; they stir the water below. Money couldn't buy or hang onto such a vantage view as ours on deck of the QE2. We stand shouldered at the rail, fascinated with the positioning by the churning craft below. We show hearty adroitness holding our rail spot. We share as participants in delight. Whether on sea, in air, or on deck, it is fun. It appears to be a naval dance dating from Minuet to Rock and Roll.

The program's highlights follow:
A Review of Embarked Veterans by
HER Majesty The Queen.
Queen Elizabeth II
and the Heads of State or Government of
Australia, Belgium, Canada,
Czech Republic, France, Greece,
Luxembourg, The Netherlands,
New Zealand, Norway, Poland, Slovakia,
and The United States of America

English Channel – D Day + 50   299

Thirty-two ships at anchor at Spit Head off Southampton. Fifteen others close by at Portsmouth. We on the QE2, first in line, anticipate the show; U.S. Navy Carrier George Washington, second, is next to us. The side-wheeler O'Brien churns by often. An ancient bi-wing torpedo plane flies agonizingly slow along the line of ships. Spitfires and Mustangs appear; their Rolls engines purr sweetly. This is D-Day plus 50 years minus one.

Today is the mammoth lead in for tomorrow's Normandy commemoration.

RULES AND REGULATIONS

QHB Queens Harbour Master Portsmouth tells what and how:

A fine blue gold trim, one-sheet folded, slick publication with map and photos sets the scene.

Regulations are listed from A through K; here is K.

k. Mariners are reminded that the expected congested conditions together with confused seas created by the wash of vessels will make the navigation of small craft particularly hazardous and demanding.

Timetable includes rules and dates June 2nd through June 5.

Times are specific on June 5: last two items are:

12:45 Fly past of veteran and modern aircraft.

12:50-13:20: HMY BRITANNIA reviews ships with veterans embarked then sails with International Flotilla to Caen.

Viewing is really something. Last night the George Washington and QE2 swung at anchor in position. We marveled at the close view of the George Washington. Now they are ready for review and to follow the Queen's royal yacht. I am on the rail with two cameras. I feel a little crowd surge and look around. My timely selected position is hemmed in and shrinking fast. Peggy stands firm on the rail under the wing of the bridge. Folks all around are double; triple anything we have seen before. Hundreds of small boats of all sizes and description rush along. Wakes foam over wakes. The "Britannia" majestically sweeps by with Queen Elizabeth and President Clinton on deck.

After the review, we are off to Normandy, docking in Cherbourg, and attending the following:

SHORE EVENTS
1400-1700 St. Mere Eglise
Paratroopers of the 82nd and 101st Airborne Divisions re-enact the airborne invasion of Normandy in June 1944.
2100-0100 The Franco-American Ball in the Queens Room.
For this gala event, our honored host for the evening is General Merrill A. McPeak, United States Air Force Chief of Staff. Our many French friends from the Normandy area are our most welcome guests.

Monday 6 June
OUR NORMANDY DAY HAS ARRIVED
0800-1100...attend the commemorative ceremony at Utah Beach.
1200-1800 Colleville Sur Mer
The American Cemetery adjacent to Omaha Beach. The major event of the trip will be the commemorative ceremony...a re-dedication to the memory of those Americans who gave their all for the sake of liberty.

Tuesday 7 June
0900-1100 Maupertus Ceremony
Cherbourg's Maupertus Airport.
Dedication of monument to The US Ninth Air Force
Edward F. MacLean, President 9th A F Association

Ed's speech is given in tandem with his daughter, Nancy. Impressive, is the audience's pleasure with Nancy's translation of Ed's words. She has a "one of us" pleasant demeanor.

Normandy liberators Medals:

Attendees are former liberated or liberating. Others are spouses and children.

We line in three ranks of eighteen. Each of us has submitted personal information attesting to our unit and our presence in Normandy's liberation. This site's Presentations are sponsored by United States 9th Air Force Association.

We receive medals one rank at a time. I am in the third rank. Presenters are mayors of Normandy cities, towns, and villages.

I am especially grateful for this medal. It is important to me. "My Mayor" can tell I am feeling emotional and serious. I am remembering my closest friend, Chris, and how easily he talked with twelve year old Ramon at A-12 and of our men lost. I have not forgotten the day we four replacements arrived. "Take 'em down to the 377th. They lost four yesterday." We knew then how green we were.

My Mayor is maybe thirty-five. He smiles in animation in an attempt to cheer me, while he pins on the cherished medal.

Yesterday, at the American Cemetery we witnessed the missing man formation. It has emotional impact. One of a flight of four jet fighters zoomed to the heavens, from directly over the sculpture, "The Spirit of Youth." The combination is as powerful as that of the Lincoln Memorial in our nation's Capitol. Here in the American Cemetery in Normandy, the Youth statue seems to represent beauty and life, then sorrow, of mothers and fathers, wives, and children.

Return 94, like 84 Return, takes planning with objective, detail and execution. These traits and talents when connected with desire result in powerful action.

The folks planning and leading 84 Return were also among the leaders, back in the late fifties, when the 362nd Fighter Group started an association. These same men (their families too) joined with all WWII Ninth Air Force commands forming a Ninth Air Force Association.

The Normandy D-Day commemoration became the premiere coming out event for us, General Patton's fledglings. I started from this beachhead ground. We dug in, flew in flak, dove for cover, shuddered under enemy flares, survived strafing, shelling, and bomb attacks.

We air-bridged over hedge rows, choke point ambushes, and cleared the air and swept the ground. We saw the need, learned, and helped.

We will be coming back to Normandy in ten days.

Now we start for our second objective, to see the scene of my parachute landing near Weidenthal Germany. Chuck and Fern Mann, friends and encouraging instigators in the aforementioned Group and Ninth Air Force successes, make up the other half of this mini sub-tour.

Fifty years after reprieve, escape still seems a gift. I am eager to see the spot in the forest where my parachute whapped me down on new-snow by wild rose bushes. Liesel, seventeen then, heard the Jug and its explosion. We will know more when we get there. She with Uwe, humanitarian businessman, and their families, with support and aid from kind people of Ramstein Air Base will show us both the parachute and crash sites.

We load excess baggage into our Normandy friends' car. We will be back in ten days, for a quiet, tasty, remarkable visit with Michel, a former French fighter pilot, and his lovely wife, Beatrice Bouvier-Muller, who is researching and instigating the placing of memorial monuments for all US Ninth Air Force Normandy air strips.

Chapter 42:
Liesel

We head out in our Cherbourg rental car; Chuck is at the wheel. Fern navigates. I hold a Triple A Road Atlas. Mid-afternoon, I take a turn at driving. The speeds seem terrific, partially because the dial measures in kilometers per hour which always seem to exceed 125 just to stay out of the way of traffic from behind. It is as though we purchase time on the toll road rather than miles.

We contribute equal amounts regularly to the money bag for fuel and tolls. Peg keeps the books. I anticipate and reminisce. We have reservations in Reims. I also have a cold.

Ten years ago in France, Peggy was badgered with a cold, too. She said she didn't have time for one then and neither do I now.

Next morning I feel good.

We drive out to Chateaux Sept-Saulx. Charles Minot, his mother and step-father are gracious. They show us the 362nd Fighter Group plaque, presented by the group, honoring those American pilots lost while fighting from airstrip A-79. Chuck and Fern had brought the plaque to Chateaux Sept-Saulx previously.

We need to move along. We drive to the WWI monument where A-79 airstrip once crossed this road now going north to a present day firing range. We eat our picnic lunch within a few yards from my last combat lift off.

We must pass up Clermont-en-Argonne, though I will never forget its stark white tribute, their monument to the one-hundred men and boys taken in retaliation and the seventy-two who did not return.

Douamont World War One Cemetery is next. We take photos of the cratered hills in recovery. The story of miserable mud and the inadvertent recovery is amazing. We use basin tillage at home. The results imitate shell holes. Here, shell holes retain water, allow dwell time and water penetration. A guiding theory might be if we "break it" using mechanical means, dozers, culverts, cuts, ditches, and cultivation, it might also be reasonable to use mechanical assistance to restore or maintain the hydrocycle.

Peggy and I walk across revitalized land. I have learned so much more about land and water since we were here ten years ago.

We speed by Verdun, and reluctantly pass up going to A-82.

We stay at Alvord, wanting to arrive at the American Cemetery early tomorrow morning. Our appointment at Ramstein Air Base is at eleven.

We recently contacted Dortha Kirkham. Aunty Dot lives in Corbett, a small town on our way to Portland. Her brother, Virgil, was the last combat loss in our squadron and is buried in St. Avold American Military Cemetery. Aunty Dot is a wonderful lady, was a WAC stationed in New Guinea. After the war, she took flying lessons, got her private license. She has recently received letters from people in the Czech Republic wanting to make contact with the family of Virgil. They have created a memorial and named a street Kirkham Lane, honoring his life.

We amend our schedule to visit the cemetery. We wear our Commemorate Ninth Air force Association jackets. Since this is the second day from the Beachhead celebration, I expect other families locating markers. We are early, hoping to locate two graves. Aunty Dot gave us another name. He seemed to mean a lot to her. We also scan the records for other 377th squadron members and find Second Lieutenant William Ort's name.

I knew Virgil briefly. In a late letter to Dad I tell of a new pilot from Corbett, Oregon. A squadron mate tells me Virgil coached the squadron basketball team, and they would have won the championship finals, played in Paris, if he had put himself in the game a couple of minutes sooner. He was a State High School All Star in Oregon.

We are helpfully chauffeured to the sites. We put a flag by Virgil's, marker and a red rose by Dot's friend's marker; his name, Leonard Carter, 101 INF 26 DIV Oregon Nov 24 1944.

We place a rose at Bill Ort's marker. He was from Hoquiam, WA. His and my names are far apart in the alphabet so even though we were classmates from Luke field, we never got acquainted until we came into the squadron. Bill was the only casualty on a rough second mission on

October 2nd 1944 (My 30th). The squadron came back with most ships damaged, some severely in flak encounters. Several were not "overnight" repairable.

The mission report, after vividly describing the loss of Ort and his ship in a strafing-run, describes three more incidents....

"*Red four was hit as he went in to strafe: his right gun bay door was blown vertical, the plane rolled over and the pilot jettisoned his canopy preparatory to bailing out. The gun bay door flattened out so he returned to base instead.*" I flew Red three: Bob Racine, Red four, flew my wing (Chapter 25, Goose Pits). More from the mission report:

"*On the way back Red leader* (Tom Beeson) *dipped down to strafe a single flak position firing at him and was caught at 1500 feet by about seventy other flack positions in the vicinity. More heavy flak over Metz. Red leader's plane was hit in the supercharger, intercooler, engine, prop, both wings, fuselage below cockpit and the horizontal stabilizer.*

Blue leader (John Hill) *received a direct hit in the engine and flew back minus three cylinders, one magneto and various other parts of his engine.*

We destroyed twenty seven trucks, eighteen light flak positions and four locomotives."

The cemetery's parking lot is still empty when we drive on.

Again we are in Germany; it is a beautiful countryside. Again I am reading names on road signs, but not nearly as apprehensively as on our earlier return, because a neat little lady, Katya, volunteered a beginners' class in German language at our community college. She was so clever using her middle school techniques to make our class fun. I, slightly apprehensive, shiver through my turn at expressing my reason for taking the class. She was a small girl in Germany during WWII. I was an American ground attack fighter pilot, who is beginning a search to find scenes and people of long ago. I took her class twice, Peggy once. We did learn some Deutsch. She bridged us to her former country. Chuck and Fern are far more experienced in language than we.

We are met at Ramstein's gate, and are escorted by Major Turner, Gerlinde Berberich, translator; Master Sgt. Vickie Andrews, about to complete her military career, and T Sgt. Mark Conrad, military historian. I listen to plans for this most eventful day but am counting on the others for details. These career people are used to guiding Congressmen, Generals, next of kin, maybe a President or Vice-President. Peggy and I have hosted people from chambers of commerce, or contingents of eighth grade students from city schools.

Today, there is a TV crew, also a persistent, pleasant lady reporter from the Kaiserslautern American paper. Major Turner introduces us to his team. Chuck and Fern seem like they are with home folks; Chuck has retired from flying in the National Guard. He and Fern both fly and both judge aerobatic competitions. We are farmers, outstanding in our field; the kitchen, the garden and the corral.

Inside the foyer at a Base mess and club, we are soon alerted, "The Frommknechts are here." I stammer brightly, "They're here!?" We are guided out and across parking where some of the Public Relations team chat with a trim lady wearing a green jacket and a man dressed like me. We wear ties. I feel uncertainty, a little fright. Gerlinde Berberich starts the introductions. We are half a Volkswagen length away; I stop, one foot forward, it eases a stiff back, also allows one a sprint-start in any direction. This time it is to respond to sought people. Frau Frommknecht is first. She is in my focus. Common sense stretches into assurance, automatic and heartfelt take over with an "ah, shucks" squeeze of Liesel's arm which advances into a genuine reach around hug. She and I had tramped in the same new snowfall a long time ago. I will know if she confirms the rows of burned holes in the parachute, or the shreds of burned gauntlets. Now, after nearly fifty years, we meet. The feeling is great.

We are guided through a cafeteria line. Gerlinde smiles; she is justified. Her sensitive ads in area papers brought seven responses. Two are key.

I am wanting to talk with each of the "finders," and to spend time with Gerlinde and Liesel. My story is still sort of in the egg, maybe about ready to crack the shell. I am anxious to learn rather than inform. I am ready for the hatching.

We talk over lunch; Liesel squeezes Peggy's arm and several times more during the day. Uwe Benkel, Kaiserslautern businessman who does humanitarian service in crash site searches, joins in the caravan of cars. He tells me, "I was pleased to probe an aircraft crash site likely free of remains."

First, we view the school building used as a hospital in 1944. Its four floors and location on a knoll among city blocks confirms my remembered shifting gears and swaying around corners near the end of the ambulance ride. The short stairs to the first floor caused the stretcher to tilt. Scenes wisp by; memories focus through age's ago sightless moments, some severe and some fine.

We drive out of the city and down a road following a stream in forest and through Weidenthal, out of town a little way, maybe two miles. We park then walk under a railroad bridge. Wild roses are in bloom ahead. We are stopping first at the parachute site. The wild roses fit.

I am enthusiastic, yet now a little cautious, wanting to confirm details. We walk under a railroad bridge; it has a barrier and symbol prohibiting vehicle traffic on this special use only road. The abutments are of stone. Stone? Liesel via Gerlinde did mention, "Edge of a quarry." Rock abutments, rock quarry? I walk through under the track. Track? I had left "the" track zooming up the side draw for close to a mile when I jumped. I could have bailed out near its ridge top. Is that why I made it? Got out on the ridge, and landed on the next valley's floor; I was too low at my last look, but zipped forward maybe three seconds, at better than six hundred feet per second while getting out. It figures.

I walk from under the bridge and look right, see a hill and railroad tunnel, same to the left. Ahead the foliage is green and waist to shoulder high and wet. I walk towards what has to be a stream about seventy yards out from the tracks. There are several wild rose blossoms showing, along the path to the stream. I'm puzzled. Behind is a railroad track, ahead is a stream. I did not step on railroad ties or over rails, or cross a creek. There was fresh snow but not more than four or five inches at the parachute. The creek would have thin if any ice. I was raised with a creek in all kinds of weather. If I had seen rails or a tunnel, I would have contrived to use them to avoid showing foot prints. I have confidence in the people who are showing me, yet wonder if they have confused me with someone else. I am sure of the valley where I was hit. I am sure of my targets and believe I can locate the canyon I turned up just before I was hit. My "missing aircraft report" gives the crash "Location <u>4 miles NW Lambrecht</u>." I have my map, marked with an arc on a four mile radius with green felt pen. I think the crash site will be close to the green line. (The crash site was on the east end of the arc.)

When I do find the crash location on a map, then on the ground, I will be oriented. My map is 1:250000 scale. While I am studying the ground layout and pondering possibilities, Chuck is talking with Liesel's son, Juergen, and looking at his local map spread over the hood of a car. His map is 1:25000 scale, ten times more detailed. His research and careful presentation are exceedingly beneficial.

Juergen is a forester and made the arrangements for us to access the crash site. He has marked the map with four designated places. One is the parachute site with X; the crash site with X over X; one gun location marked with a small square, and the other gun location marked with ovals. The two gun locations are on small ridges nosing down to the creek. The parachute site is between the creek and the railroad track.

There has to be a revelation, somehow. I am anxious to see the crash site. We need time with the map. We did not walk up through the quarry; it was pointed out as maybe eighty yards beyond and through very wet foliage. For the moment, the railroad has me stumped. I felt fairly isolated that day. I thought the road or wagon track my feet landed on was probably coming out of back country timber land beyond any railroad. I am sure of my targets, two locomotives, parked side by side on double track, under an overpass crossing to a village. None of my target markers are here. I have to learn more.

Peg and I ride with Mark to the crash site. He worked with Uwe and Juergen and his son in finding pieces of the Thunderbolt. We drive down the highway a half mile, and cross to a forest road. It is smooth, looks to be grass on dirt, seldom used, built on a slightly rising grade around the hill to just above the parachute site, then it takes a turn up the draw for half mile to get to the crash site. We spend time here.

There is interviewing with the video news team. Uwe has his little daughter, Dominique, and her grandparents with him. Dominique is a darling, three or four years old. She holds a small model of a Republic P-47 fighter. Uwe has done discovery and research for nearly fifty WWII aircraft crash sites. His wife is in Ramstein's billeting office. We met her at lunch.

Uwe brings out a cardboard box containing a half bushel of gleanings from the crash site. We select identifiable small items to bring home, a chunk of engine crank case, a trim tab knob, part of an elevator arm, hose coupler, a section of control gear and a very small exterior aluminum shield. Away from cameras, he shows us a paper bag with several intact but mangled 50 caliber cartridges from my airplane. The event here has become personalized.

Liesel had shown the team the crash site and Heinz took pictures. They said they made three trips to the location and spent about fifteen hours on site using metal detectors and garden tools. The crash had been salvaged decades before. Heinz took lots of pictures of the project and several of the beautiful valley and its towns. Some of the pictures were from elevations showing pilot's eye views; however none resembled the village or field and side canyon of my "target's" valley.

I ask Uwe, "Where is or was the bridge and double track, and locomotives?" His answers, "There was no bridge. Your target was tunnels."

I let his answer lay for the present.

The answer is simple. We were thinking of different valleys. We are each correct within the limits of knowledge. Uwe's conclusion assumes my plane came in view over the ridge, and gunners guarding tunnels shot it down; the pilot's parachute hit the ground by a tunnel. The airplane crashed a half mile beyond the parachute landing site.

My story is as I saw and lived it. I also made a conclusion from what I saw. My shot down story begins after my second pass to get two locomotives sitting side by side on double track under a bridge, in a village. I stayed very low and very fast heading directly towards an unseen gun emplacement. I turned hard right at the first canyon and glimpsed two helmeted gunners directly under my vertical banking wing tip. A nano second later I felt two hits, thump, thump. Faint smoke, full oxygen, instruments check, heavy smoke, very low, toggle canopy open an inch, fire rages, jettison canopy, jerk stick, release seatbelt, jam feet down to leap high, sucked out over tail, grab D-ring, chute opens, terrific jerk, ship blows up ahead, tree tops block view, look down, wow, hit in snow on forest two track road. Dropping to the second valley floor, saved my life.

Recall gives close estimates of time, together with speed, allows estimating distance. Juergen's map gives confirming distances and pulls mine and their segments of the series of events into a complete story.

Just as I did not know of the second valley, they did not know of my episode in the target valley. Too, my view was interrupted from time of cockpit fire until my chute opened just above parachute site.

My ship became visible to Uwe's gunners only after I cleared the cockpit; my figure would have been a small ball of something, smoking and tumbling, falling behind an enemy aircraft roaring from over the ridge toward them.

Before I wrote my first letter outlining my quest to Ramstein, I located a landscape and setting resembling my targets' valley having a railroad alongside a stream, with forested hills and a side canyon. Once, I had flown up this valley to get to the town where our veterinarian's clinic is located.

I drive over and find a Forest Service road rising to a place, from which a ground attack pilot might size up a target situation.

I park, get comfortable in the passenger's seat of our rugged camping van, bought used, to ease us through withdrawal from both flying and ranching. It fits well on mountain roads and under one dollar a gallon gasoline.

I want to explore the detail of the event of November 17th 1944 with recall and a sketch that would help find the crash site. I manage the sketch OK. I got much more. A very personal, traumatic reliving of a scene long shelved. In two hours, I completed two sketches, one for each pass. I sent copies of the sketches to Ramstein. Putting off fully reliving my bailout episode since Kaiserslautern hospital time, contributed to a release of unexpected emotion while sketching.

An Ex-Prisoner of War Protocol examination by the VA, forty years after the bailout experience, was completed by a team consisting of a physician, a social worker, and a psychiatrist. It was almost like, "Where have you been all these years?" The exam was fine, very complete two days of tests, exams, and consultations; they turned up some chest and eye situations of which I was unaware, as well as revealing a world for veterans I didn't know about. The reason for including this narration follows:

I almost blew my stack when I read one of the reports which described my aircraft bailout. It said, "Subject ejected from airplane."

That was it. My thought was--I owe it to my body to fully detail, bailing out at 450 plus mph with head, hands, and legs on fire, with the parachute opening at tree tops. It jerked my knees and jaw away from sockets. My escape kit ripped off; dog tags broke away; stockman's pocket knife thrust out through pants pocket; incendiary shell penetrated my seat-pack parachute, lodging spent and burning against my thigh. Saucer size burned holes were in parallel rows on my chute laying spread across wild rose bushes; my legs jammed on impact.

I really appreciate the old body hanging on to all the loose joints. "Eject" was the word to describe getting rid of the canopy in 1944. "Ejection seats" were developed by Doctor Stapp at Edwards Air Base after the war.

Mark and Uwe remark how accurate my sketch is including the anti aircraft guns. Their remarks lead me to some mixed conclusions. They had talked with a man who was operating one of the guns shooting at my plane and thought had brought it down. Now, I have their story and mine too. Both use my map which actually describes the target valley. I believe my sketch is detailed and accurate. I know I will recognize the target spot when I see it. I feel energized and try for calm. My cold is becoming a doozy. I am running on adrenalin with no drip. I just put this information on hold.

We drive to the Weidenthal City Hall. The walk from the car to the entrance seems slightly down hill, about right. Inside, I expect to find long stairs, straight ahead up and not quite as steep as those in the ranch house I grew up in, but longer; instead, these go ninety degrees left to a landing and then another turn with a short stairs to a foyer. The council room is nice with chairs and tables. I am introduced to Obersburgermeister Niederbergger who firmly shakes hands and holds his grip through several spoken sentences as he presents me with a fine book* about his city and history of the area; He mentions my "over night" stay in their city although I didn't pick it up until we listen to Chuck's video tape. I wasn't sure of his words. Gerlinde is thinking I understand way more than I do.

The mayor presides, wine goblets appear, corks pop. We are toasted in the room of my interrogation. I saw no folding chairs today.

We exchange gifts with the Frommknechts. Peggy gives Liesel a century old American Whitehouse Cookbook printed in German language. They give me a ram's head stein and Peggy a gold trimmed china plate, cup and saucer. Liesel and Heinz also give us a recent photo book presenting villages of the area with English sub text. They have each signed the book: *For David and Peggy Childs from Liesel Frommknecht and Heinz Frommknecht.*

The video team asked questions especially of Peggy. Peggy responded. I listen, learned. We didn't see the news clip. The reporter and I understood Peg's words and her demeanor in answering, "…with faith."

Uwe pointed out a much younger man than I, saying something like you asked if anyone remembers seeing you. "He was a little boy when he saw them carry you out the next morning. Do you wish to meet him?"

No mistake, "Next morning." At first I'm befuddled, makes no sense. I think again, maybe I am not their man. I need to straighten out facts from conclusion. I indicate being unsure and was interrupted before completing my thought. I regret not following up; I wanted to verify one's presence at the time when I was carried from the building. Recall of the "footbridge" incident would confirm presence.

Going out of city hall, I ask to cross the road and walk downhill towards the stream. A red car is parked in a vacant area which once held the little building where I was kept for an hour in the afternoon of November 17, 1944. They say, yes, there was a building there; there is also a foot bridge, though different. The former building was for crew and passengers. The wooden bridge was to cross up to the tracks. I never saw up to the tracks back then. Certainly anyone spending the night in the building would have heard the rumbling of trains after nightfall. I heard no trains, whistles or engines.

We ask Gerlinde to arrange a dinner locally. Peg and I host the small affair for all those enabling the fine results. We are grateful. We really hadn't had time for visiting. Dinner is fine. We have pictures of the dedicated crew and the hospitality of the restaurant. They presented us with a bottle of wine and also one with the date and occasion painted on it.

My cold needs a rest. We need to regroup. I need a hot bath and time to list the unknowns, account for differences in my story and the one developed by the Ramstein/Weidenthal team. The score is the crash site and the parachute site are in the right relationship to each other. They fit. But, the railroad track and creek need some deduction; the staircase is not straight ahead, and I didn't stay all night. The man was a little boy fifty years ago. He may have seen me carried out and placed standing on the foot bridge. Being a little boy and knowing what little boys do first thing in the morning, he used good logic. Actually, he saw me doing what fighter pilots do after a mission. The lad's time of day reasoning, same as mine, depended a lot on fifty year old facts. He related to back then. He may have also remembered my open and swollen burned face with no eyebrows or lids showing as I was walked from the bridge to the car. I departed as I arrived, with no medical attention until a few miles later with the medic I referred to as "General Goring."

It seems clear, I was hit and out of the plane before it came into view from over the ridge between two valleys, Hoch Speyerbach and Speyerbach. It is likely my airplane, with empty cockpit, passes over the tunnel while ack ack shells fire at the already fatally hit ship. Its bombs and bullets have struck their target 15 to 17 seconds before; the plane likely is out of trim from increased speed and loss of cargo. My target was nearly two miles behind. I was shot down by a gun crew back on the other side of the ridge. I moved from valley to valley but at the right second. I am lucky, vortices put spin on my body so centrifugal force slowed my reaching the 'D' ring to just the right instant to land me on the narrow second valley floor. Earlier or later, the terrain would have met me before the chute blossomed. The airplane, with empty cockpit, passes over the tunnel while ack ack blazes at its fuselage. I was downed by the gun crew back in the first valley. I need time with the map and to find out more.

Snow frosted wild rose hips showed on bushes by my parachute. Wild roses in bloom and the distance between the parachute site and the crash site do make sense.

"Weidenthal" History of a forest Village by Heinrich Stuckert, Forest Engineer. The book is impressively researched and details political, religious, and industrial history going back to the fifteenth century and includes both my target and my parachute valleys.

Chapter 43:
Next Day

Our Ramstein visitor quarters would please a congressman or a general. We are happy as two clams in a just right tide. The decor is attractive, and fridge is well stocked. A hot shower eases my cold symptoms.

My darling suggests a beer would be good for the cold. We relax and analyze our day. My first exposure to such elegance in Germany was under guard, being escorted through a couple of first class passenger cars up to a luxuriant club car the night of February 20, 1945.

Liesel, the Ramstein folks, the citizens; everyone in the Thunderbolt discovery team are vivid. I took dozens of photos: Chuck did plenty of video. Fern, too, took lots of film; Peg kept her notes.

This day, beginning with the American Cemetery at St. Avold, seeing the chute and crash sites, Weidenthal city hall, wooden foot bridge; all, including France to Germany, will yield years of reflection. We study Juergen's map, measure distances and find close to probable. We store our thoughts until tomorrow's light.

Our evening deductions do fit. I walked around the rose bushes and uphill from my parachute, opposite from the creek; didn't know it was there. My steps went toward and up the hill and found no railroad track because it was beneath me in the tunnel. My walk ended above and on top of the tunnel. My captors collected their souvenirs from me, while we stood above the tunnel.

Yesterday, Ada Martin asked us to meet for breakfast at the base's Woodlawn Golf Club. We make up for yesterday's intense learning while with new friends. Most unusual, for me, I left an unfinished plate. Not this morning. Ada was patient then. She has our attention today; we eat and answer questions. We had orange juice in our quarters so we calorie on with a great breakfast. Peggy orders Woodlawn Scramble; ham, potatoes, with whole wheat toast and coffee, honoring home when we ranched. I have the same, only with rye toast, a mild reminder of prison camp.

We are looking forward to the next ten days; enjoying today and another in Germany, five more in France through the Loire river valley, then three nights and two days with Beatrice and Michel in Normandy. These are "vacation days;" they will be free lance soon as we tie up my part in the attack and bailout part of the story.

Fern drives; Chuck is in front. I sit behind Fern, Peggy beside me. Peggy buys highway miles; Chuck keeps us to a very elastic schedule; I watch landscape.

Weidenthal is on our way. We drive slowly, enjoying the little city; its Maypole is wound and still standing. We walk by shops around the square, we don't stay long. On down the road we stop briefly at the parachute site. We stop at another bridge and tunnel. I walk up onto the track. Chuck asks, "Anything up there?" I answer, "Nothing."

We drive through Neidenfels, Liesel and Heinz's village. Peggy and I are each acquainted with small towns with railroad traffic and railroad families.

We drive to Lambrecht, the town identified in the Missing Aircraft Report. I have little recollection of Lambrecht's orientation. My concentration focused on two locomotives, side by side, nosed under an overpass leading to a village from the hillside and crossing the small valley. My calculated "best fit" is Frankeneck. We turn back; in less than a quarter mile, we see the two towns blend together joining the brook, Hochspeyerbach, with a larger stream, Speyerbach.

We turn towards Frankeneck along Speyerbach; choose a side street rising to a road looking down over town. We look upstream. It follows alongside a single track. "Not there," I say. Chuck is following the track back downstream with video camera. He says, "There it is." Peggy's notes read--*sighted, two track, three track, there it is*! We are looking down through tree branches on three pairs of rails. Two look old. We drive down to the main street and find a foot bridge at the rear of an official looking building covered with vines. The foot bridge leads across Speyerbach. Here it is a beautiful narrow, deep flowing stream with stable banks. The path up to the single track leads between two fine well-kept vegetable gardens. Chuck follows me and waits by the gardens. I turn left downstream, maybe two blocks, before I find double tracks. The ties are old and cracked. Now what about the overpass bridge? I walk a little further. There lying against the toe of the hill is a large broken chunk of concrete. It looks old. I take photos. The hillside is stabilized by old rock work firmly imbedded on the slope up to about the height of the overpass as I remember. An overgrown berm wide enough for a smooth roadbed. It leads upstream on the contour along the hillside above the railroad tracks. I follow it briefly. I think of men and horsedrawn construction. It is distinct with established grade, low edges, minimum cut or fill. Trees and brush grow alongside and almost over it but do not obscure the formed roadway. I climb down and go back to the gardens.

Chuck is interested in a lot of things. He can see a mood change even though I am trying not to reveal one. He is taking it easy, looking at the gardens, and wire fences denoting great care. He opens with, "Did you ever see such tall bean stakes?" The stakes look to me like tepee poles used by Indian families along streams in our mountains in summer when I was a kid. His next question comes the same way, "You found something, right?"

I answer, "Right!"

We cross the foot bridge and join Fern and Peggy. It is noon time but our great breakfast is good for another hour. I am on trail; so is everyone. I probably had repeated myself recently. Like many, Chuck hears what and when he wants to; then he hears and retains, even ponders the scene. I take pictures of the ivy covered building. Fern starts up the road towards where I had banked over in a hard right turn, about a half mile. Chuck narrates the scene of a hard hitting Thunderbolt hugging the deck, moving out, "I mean moving! Everything full bore!"

For the moment I am there again; I feel its zoom speed compounded with full throttle, my elation, skimming the creek side meadow. It was only about three seconds then. Fern stops at the canyon coming in from the right. It is here I glimpse the gunners, then felt two hits. I was heading for unknown. I had used maybe four seconds; it was over in a dozen more.

We are at the side canyon. Things look pretty much as I expect. I am certain of this place in relation to the double track and now I want to confirm the abandoned road paralleling the railroad coming up from Frankeneck. Fern drives near the left or railroad side of the main canyon.

Next Day

I have three cameras, two are auto, and one is manual advance and rewind. I am good with operating all three. I am only seven decades along and should be bright; am really happy with the situation. I stride off, crossing the railroad track, eager, anticipating intersecting the old road bed. First though, I stop to shoot pictures back across the valley toward my escape canyon. I let the camera down on its shoulder strap next to the other camera; it starts whirring in rewind. I hurry on, undoing a new roll of film, then grab a camera and slide the opening release latch. I am really cooking. Not stopping, I charge on, grab for the used rewound roll, and instead find a roll half exposed and a quarter ruined. I save what I can and reload both cameras; the one really out of film and the one fouled up. Not wanting to retake or use extra time, I bear up to being fallible. I chug on and to my extreme delight; find the old overgrown road bed extends upstream by here. I speculate its long ago purpose; mining, agriculture, lumber, firewood?* Steam or horse drawn. Rails? Wood, or iron, or steel? The berm seems to have constant grade. I think of following it for the kilometer back to the long-gone overpass, which slightly hid my targets on November 17th 1944. Instead, I follow it for a hundred yards, and then satisfied with my return to Frankeneck, I am ready for lunch. So is everyone else.

Frankeneck *Gastehaus* is on the right as we come back to town. Lunch is pleasant. I remember the shepherd dog and the window boxes.

Peggy writes: 4 pm we are in Lindenberg for —*"Tea time"- strawberry ice cream, a wafer and coffee--geraniums--Onion dome steeple*--sums our day's three village visits by adding one word--*Lovely!*

We share the dining room with another table of four who are conversing pleasantly in Deutsch before a large picture window. The ambience is comfortable after my recent intensity. Paintings or pictures hang on the wall. The picture window shows timbered hills rising beyond roof tops.

We decide on heading towards Speyer but hope to find lodging on the way. We have three hours of daylight. My eyes are captured en route to the doorway by a large painting on the back wall. I cannot believe it. I am totally surprised. The landscape is of my attack scene centuries ago. I have the image, forever, of these three knobs. I rolled out of a left turn to parallel them in an attacking swoop on two locomotives. I sized it up from above the knobs prior to each of two attacks.

The painting is large, shows three knobs, and the north side of the valley in summer greens and yellows. A dirt road crosses a wooden bridge over Speyerbach and leads uphill passing two or three red roofed farm buildings on its way to the castle dominating the first knob. I used a telephoto lens this morning for a photo of the castle's centuries old ruins. There are no railroads in the painting. Much of the timber has been consumed but lots of small trees are growing in stands between small cleared fields. Hochspeyerbach enters from the right, beyond six or seven tall evergreens; the trunks are clean to the first limbs twenty feet up. The painting is such a surprise, I can't compute.

It certainly depicts a long ago scene. I should have learned more, much more. We walk down short flower decorated steps to the street and go on our way.

I will always remember the knobs--hopping over the overpass--skimming the little riparian field; full throttle--bank hard right--a gunner's look, so close, so quick his lips trailed his open mouth--.

We received an envelope with a Neidenfels postmark, in November 1994. Liesel and Heinz sent a beautifully done 50th birthday card in the form of a birth certificate, in an ornate scroll. It gives the date 17th November 1994.

Written on the reverse are the words:
God give the second life to you in 17 Nov. 1944
Heinz & Liesel

* Friends, Inge and Norman, translated parts of the 'Weidenthal' book finding it mentions Frankeneck 1832, paper and plank mill; 1855 washing, cutting and grinding...of the needs of the paper mill in Frankeneck, i.e. paper pulp...It seems the road and bridge preceded the rails.

Chapter 44:
Rhineland to Normandy

My parachute dropped me within fifty steps of Hochspeyerbach, which joins Speyerbach just below Frankeneck. I first heard its trickling waters fifty years ago in Weidenthal, upstream from the parachute site, when two Wehrmacht guards carried me from a temporary holding room and stood me on the deck of a wooden foot bridge. I could hear but was not seeing much. (Chapter Two "Fender")

The German guard directed I blend "Allied" waters with those flowing to the Rhine. Speyerbach flows to Speyer and empties into the Rhine. My tributary "incident" was probably of lesser importance than General George S. Patton's "caper" when he crossed the Rhine three months later.

We drive to Speyer and visit Speyer Cathedral. Red sandstone, and decades of labor and the skill of craftsmen and artists combine with medieval engineering and architecture to create height, scale, and enduring strength. Its age and happenings are impressive, however, even more impressive, it was constructed in only thirty years.

Next, we exclaim over magnificent cars and airplanes in Technik Museum. I cannot relate with the ejection seat exhibited. We didn't have them. Apparently, this one succeeded and was used by German pilots in WWII experimental and jet combat aircraft. It is displayed with a manikin wearing a protective face guard. Its limbs are held in position by paddle like retainers and fully protective clothing.

A Russian Mig 23 is prominent for an up close look. I enjoy sketches of German WWII fighter planes and pilots, done with a little buffoonery, not unlike those of Bob Stevens' "There I Was" fame for poking fun at U. S. pilots. One sketch shows a pilot leaving his parked FW190 with his riding crop under his arm, striding in full swagger, wearing cowboy spurs, with outsized spiked rowels.

We cross the Rhine over an impressive postwar suspension bridge. I snap a photo as we go under and by the center tower with its web of suspending cables.

We lunch in Ludwighausen, *"mushroom soup, and wiener schnitzel."* Heidelberg is special for Chuck and Fern. They alert us for the train station, where hundreds of commuter's bicycles and the daily entanglement of handle bars, seats, and wheels they marveled at twenty years before. It hasn't changed!

We cross Old Stone Bridge and up the winding road to Konigstuhl, King's chair, and ride the elevator. We are at "armed recce" (reconnaissance) altitude looking across forest lands into clear distances then down on the Neckar river. This scene, city below and beyond, feels to be a rich and beautiful land, showing care and an expanse of forest and agricultural resources.

Again I wonder why civilizations must pull and tug and battle for more. Climate, soil, water, weather are tools of food supply. Roads, energy, mobility, government, military, housing, safe working and living surroundings are not for everyone. Will they become so? Civilizations seem to not civilize.

Worldwide, man has managed survival within abundant adversity; Arctic, jungle, steppe, island, coast, desert, and in rarified air of very high mountains; man has adapted. We need learn to adapt to each other.

These prosperous cities, Speyer and Heidelberg, are not large. Heidelberg, now at 150,000, and Speyer near 50,000 were mostly spared in WWII. Heidelberg lost its Old Bridge but rebuilt it soon after and it remains "Old Bridge." Speyer's bridge over the Rhine was destroyed in 1945. The one we described earlier was constructed in the 1970's. These cities have felt strife and change from repeated conflict through ages. Now it has been a half century since a hostile crossing of the Rhine. Is this a record we can build on?

Back in France, we overnight in Besancon. Next morning, we breakfast in the train station. Our talk is planning from here to Normandy. The Loire river valley and some of its chateaus is our focus. We have the fun-to-travel-feel. A chateau is up on a tree lined ridge that slopes gently away both left and right. Our view to the "castle" is steep and across open farmland. Its close border is a tree lined stream with fine cover. I think beef cattle, game birds and a dog; I bet Peggy is picturing wild flowers, song birds, and a black mare.

Auxiere takes cash nearly on the go. At St Fergere, Peggy's ancestry simmers to the top. Her menu is *–Peach Liqueur, Terrine of Pate' de foie gras,* (goose liver to me).

Orleans is our daylight destination. Gien attracts us; Fern stops under a shade tree. Our small town and country beginnings tell us we are just in time for a home town parade. A commanding lady, standing on a platform in the middle of the square, holds a microphone. People, teachers, and kids are in pre-parade doings. A John Deere Tractor passes by pulling a flat bed trailer with hay bale seats. A squad of teenage young men in ties and white shirts are ready. A red tractor pulls a trailer with an enormous horn of plenty. We are now in the valley of the Loire. School seems to be over. It is time to finish work and have fun. We love it!

The lady with the mike addresses the situation with drill sergeant's cadence. She speaks with such clarity; I wish I had a duty. I could do it, even though I cannot understand a word. The parade starts. An enormous red fish extends beyond both ends of a blue trailer; little girls sit around the Horn of Plenty; a dozen little boys are on decorated bicycles. Another trailer joins up toting a one-fourth scale model Dutch windmill in red and yellow flowers and buntings. The sails reach above two accordion players and a dozen young riders. Everyone with everything falls into line and moves out of the square and down the street. The dedicated are under the guidance of the most dedicated.

We bid adieu to home town reminiscing. Shortly, we are in Les Bordes and wow! I'm suddenly back in fourth grade geography, looking across a city block to a genuine Dutch, though French, wind-driven flour mill. Fern circles back and we stop for a look. Not knowing about windmills other than for ranch wells at home, I am cautious around this one with its low sweeping sails. They are still now and seem to be safely locked with brake set. I walk around peering for understanding. On my way back to the car, a most informative young woman appears from inside, with greetings. I am over my head after her, *"Bonjour, monsieur."*

I say hello and beckon for Chuck. Madame and her husband maintain the mill. She also markets her embroidery. Peg bought a piece.

In a short time, with encouragement and a few questions, we are up in the tower's innards where Madame uses a steel pick to demonstrate roughing up the grinding face of a millstone. This leads to releasing the brake and the sails turn in the light breeze. I know enough about cleaning, moving, grinding, and sifting grain to tell this is a development from centuries past, yet still doing a job with only the forces of muscle, wind, and gravity. Crum's old mill in our community used water power; it quit operating about twenty years before I was born. Our cheerful guide moves flat belts from idler to drive pulleys, adjusts screens, tests tension on belts and checks the throw of shaker arms. The whole three stories of motion could thrust one into a heel and toe jig. I probably would have tried a little of "Turkey in the Straw" but it all looked so fragile I might have literally brought down the house. The mill is long on being amazing and pleasant.

The Loire valley is still target familiar. It was the right flank for General Patton's liberating Third Army's drive across France. XIX Tactical Command, under Air Force General Weyland protected Patton's flank and constantly whacked down on enemy units coming into the Loire. Our 362nd was one of General Weyland's ground attack Fighter Groups. We did such a fine job, a German General, a Division's Commander, said he would surrender but insisted he would do so only if a P-47 Fighter officer was present. In our words, "he surrendered twenty-thousand men if we would call off the Thunderbolts." Hearing this news, we slaunch our crusher hats another notch.

It has been only five days since our Normandy D Day celebration. Hundreds if not thousands of participants are spreading through European combat and occupation zones. People are very kind to us. The Le Rivage, is a semi-resort hotel. We check in then meet back in the ground floor lobby. Well groomed grounds slope down to a canal beckoning us to "look see."

We agree, "Let's walk." Before we get started, our host, M. Beraud, greets us with a non-refusing invitation. We follow him to an out of traffic place--*for a small Champagne soirée with delightful petit fours*. There is more. He suggests a toast best remembered as, "To American

Aviators and to our Liberation." Peggy and I know a little of the feel of liberation.

We walk along the canal; snap photos of a flotilla of ten recently hatched ducklings paddling in the wake of their mallard mother. They are not quite showing the discipline of oarsmen in a racing shell.

Next morning at a gas station, distance travelled is two thousand and ninety-six kilometers or 1300 miles. In Orleans we walk around a heroic statue of Jeanne d'Arc mounted on her war horse, wearing armor with sword in hand. She was a major person in my grade school history discoveries. I saw her as one very brave person.

A brochure found upon entering the Cathedral of Sainte Croix cautions, *"In a museum, a certain decorum is desirable. Much the more so in a church; please be careful."* In the beginning of our visit, I am slightly amused at the translation. Before we finished seeing a lot more, I am thinking, "Very well done, entirely appropriate." It is remarkable how one's point of view will change with extra knowledge.

-Not to break the silence, and, if you are a believer, to turn your heart towards God.

-To abstain from eating or smoking...

-Not to go into the church with your dog...

-Not to circulate during the celebrations.

This cathedral, built during the thirteenth and fourteenth centuries, gives much interesting history including the following:

Joan of Arc came into this cathedral on May 8th 1429, and participated in the first procession of thanks for the deliverance of the town. For more than five centuries, every year, on the same day, a commemorative procession renews this action of Joan.

Elsewhere we found a twentieth century memorial plaque bearing the following inscription in English above a similar one in French.

UNITED STATES OF AMERICA
IN REVERENT MEMORY
OF THE MORE THAN
ONE HALF MILLION
AMERICAN FIGHTING MEN
WHO GAVE THEIR LIVES
FOR GOD, COUNTRY, AND FREEDOM

DURING THE TWO WORLD WARS
1914-1918 AND 1939-1945
AND OF WHOM 67581 REMAIN
IN THE SOIL OF FRANCE
* * * * *

Chambord Castle left Peggy "note-less." She reacted to Chambord like my little friend, Cindy dog, upon seeing her first peacock in full strut...Impossible! The grandeur and mammoth scale on an open plain has elements of Stonehenge only because each is on a grass covered plain; both used artisans and engineers of their time and place, with energy of their day. One is oriented to the universe and celestial, the other to leisure with excess. My interest is great and knowledge minute. By the time we carry on, my brain, legs, and film are exhausted. At any rate we enjoy the scenes for almost three hours and end with adventurous chestnut ice-cream cones.

Chinon area seems so interesting we've scheduled two nights here. Fern and Chuck navigate skillfully into a very narrow street with half on sidewalk parking. Cars maneuver deftly; fenders miss and merge with audible breaths taken in and whistled out.

We check in. The host is expecting us; before sending us upstairs with a guide, he says confidentially, "We have a little surprise for you. Would you come down at eight-thirty?" Chuck, always one for clarity, asks for confirmation. Our host responds directly with a twinkle, *"S'il vows plait, Monsieur, precisely."*

Peg's and my portion of this late medieval, four level plus attic, structure, with a spiral stairway tower, is on the third floor with two slightly tilted steps up into our room. Chuck and I synchronize our watches at the stair landing.

The view from our four pane square window is over a few roof tops and up to Chinon Castle stretching along the top of a sheer bluff. Tomorrow, I will make a first thing reconnaissance hike up there. I put it aside right now and take a Chambord induced nap.

Refreshed and curious, we meet at the spiral staircase a trifle early and plan our descent to appear on final approach "precisely" on time. Monsieur and Madame greet us and escort we four from above through an ancient arched doorway onto a beautiful outdoor terrace with flower decorated tables. Pink climbing roses arch over a wall's doorway. Other guests, employees, waitress, waiter and Chef, all seem genuinely sincere in greeting us with standing applause. We are wearing our Ninth Air Force jackets. Chuck's and mine are white with blue trim. Our sweethearts' jackets are white with yellow trim. When vintage champagne is in everyone's glass, our host offers a toast which includes the words, "thank you" and "liberators." We are moved and feel welcome. All are warm and most converse in our language. We engage in small discussions during and after dinner.

The Menu: *David and Chuck salmon, Peggy medallions of lamb, Fern veal...red wine, fromage tray...Crème Brule.*

Because of my close ties with two fellow prisoners of war friends, British Jack Lomath, a French trained Chef with a career in culinary arts, and Richard, also an ex-Thunderbolt pilot who learned from Julia Child while on military assignment in Oslo, I am totally, I wish, prepared for this atmosphere and fine cuisine.

Morning is quiet. The car squeezed with two wheels on curb is unblemished. I set out for the castle using third floor window memory. I start by moving up until blocked, then go right until unblocked until I reach the top of the bluff, then move left. The path is slightly downhill. I come out on a small flat bordered by low stone walls. A deep, always dry, moat is below. It displays replicas of siege weapons older than cannons. One is a giant bow, another a boulder launcher; items I might expect in Prince Valiant of the Sunday comics. Hunger trumps curiosity. I don't wish a chance meeting with Aleta so I hurry down to report the potential and ease of access.

After breakfast, we are on our way to–Chateau d'Azay-l-Rideau.

We drive a short half hour through farming countryside. The Chateau is symmetrical and petite as compared with Chambord. I like the tree lined walk to the building and direct access to its entrance. The outside beckons me for a walk around. It appears to rise from the earth on the front and from the water on the left and rear.

We enjoy driving with stops to take pictures and snap some on the go. Surprises pop up including a nuclear power plant with cooling tower, a truck loaded with saw logs, like found on western Oregon highways; then a single pier in the middle of the Loire--looks like a sacrificed bridge while protecting Third Army in our war. Farther, in our journey, I see a scene from my youth. Fern stops and backs up to where a crude model airplane is mounted on a pole, away from the road. My glimpse at road speed, saw it tucked between two red roofed buildings with a camouflaging back ground; I thought B-26 Marauder. Nostalgia sets in. I remember my own models of older planes carved from cedar fence posts. I left one on top of the corral gatepost.

Up close, it looks more like a plane used by Admiral Byrd to fly over the North Pole exploring the Arctic in May 1926. No wonder it's cracked and weathered wood and placement of fixed gear look more WWI than WWII. I take two pictures. The telephoto shot shows the vertical tail of old and rusted galvanized sheet metal, the crude work of a young aspiring aviator of long ago, most likely after WWI. My last "home built" plane on a pole was with retracted gear and ten years later.

Back to Chinon and to The Castle. My early morning foray to its ramparts had given me only a scant sense of what was there. This time, I discover the huge, maybe fifty cm diameter, round stone ammunition for the flinger. This excursion with friends is far reaching, all the way back to Joan of Arc's time, and up to the top of the clock tower. We are intrigued by maps showing details of her Kingdom to Kingdom sweep to the east and north. A visitor's guide gives dates--Joan came here March 9th 1429, left April 20th, entered Orleans 29th April. A wall map shows continuing campaigns. An abbreviated list of cities taken gives Tours before Orleans then Gien, Auxerre, Troyes, Sept-Saulx, and Reims.

The route is especially interesting to us because we follow parts of her trail that ends in Reims. Our fighter strip, A-79 was east of Reims twenty km near Prosnes. Our quarters were in Chateau de Sept-Saulx.

Chinon Castle is part museum. An early clock with huge bronze gears, mounted in an open frame is in contrast to an iron mechanism which might be described as a forge fitted iron panty. I'm told it is a chastity device. I am puzzled enough to ask only myself, the obvious. Are these objects significant enough to differentiate between the Bronze Age and the Iron Age. We rest and refresh before our second dinner on the terrace.

Breast of Duck
Mushroom Poached Pear
Fruit and Sorbet
(Kiwi-Strawberries-Bananas-Orange-Grapefruit)

Rose' Wine

Leaving Chinon 7: 30 next morning short distances, plenty of time, and looking forward to days in Normandy with Beatrice and Michel has us in best of moods. This is a lots of "Kodak" day. The first photo shows a very slim little lady in long black tube-like dress and black scarf. She is leaning forward jawing a man on a motorcycle in a crosswalk at a stop light. Next, image is a street level window with intricate lace curtains, wide shutters and a black cat sunning itself on the outside ledge. Down the street or in the next town, a beautiful orange marmalade cat sits majestically against a stone wall by a curb's step.

We stop at Ambrieres-les-Vallees. A very tall sharp spire thrusts skyward, a note on the back of my photo pleads, "Don't force me to parachute here —"

We are heading north with our afternoon goal in easy reach when we are attracted to the city of Flers by its banner across the front of a white gable end of a large building. Polish and US flags are at bottom left and British and Canadian flags at bottom right.

1944 – 1994
WELCOME
TO FLERS
TO OUR LIBERATORS

We unanimously overcome our "shy" tendencies and park in a spot near a street fair and open market. People are bantering and bargaining at booths, two rows down the middle and one on either side, and the adjoining church commons. We listen, walk, and enjoy seeing the full range of offerings, finally meeting centrally to compare notes and decide where to eat.

We feel welcome in Flers, so much so, we dine on *Croque-monsieur* – toasted ham and cheese, and *Croque-madame*--toasted cheese with one fried egg on top. We also have O*rangarina*--orange drink in a bottle, coffee and an apple tart.

The street fair borders a very large stone church with a rose window. On our left, facing the church front, is an American flag on a pole flying over the booths. Eight wires bearing Allied flags span the fair and brighten the church, Eglise St. Germaine.

Driving through Vieres, a veterinarian's office displays a cartoon portrait of a long-lashed, young cow, wearing a cowbell strapped to her satin neck.

The cities and towns have dressed brightly with flowers planted in special designs. Vieres' exhibit is a floral low-wing fighter plane with inline engine.

Chapter 45:
Beatrice's Normandy

Isigny, Normandy: We first met Beatrice when she was a guest of the Ninth Air Force Association in Colorado Springs last year, May 1993. She is a bundle of enthusiasm and a mover of people. It is great fun to see her marvelous Norman home and meet family. A son, Francois, and his friend, Sylvie, visiting from Paris, greet us. Chuck and Fern have visited Beatrice and her husband, Michel, before. We hope to share Michel's early morning mission to the bakery for fresh pastry and bread.

We admire the stone buildings of their enclosure; barn and shed where horse training and driving gear seem to be in retirement. Beatrice's office is across the courtyard from their home. This building is long. Each room has an outside entrance door.

Their house, too, is long, has two stories and a usable attic. Ours is second room to the left beyond the stairwell. Doors open room to room without hallway. We pass smooth stone stairs on way to our room; steps look worn, with centuries of use and scrubbing. Fern tells us not to pass an invitation to visit the attic.

François is in the middle of a tough mowing job using a John Deere rotary mower in tall pasture grass on the hillside below the house. He needs more horses, either to graze or as horsepower. I am more valuable as sympathizer than consultant. We are happy when called to join the others, now all in the backyard.

Beatrice has the fixings for supper. Michel is jocular and is a French Navy pilot with entertaining skills. We welcome a walk. The whole troop, save Beatrice, moves out past the flags of our two countries, flying from twin poles in the courtyard. The board gate is open since there are no longer any horses. Beyond is a beautiful stone church with buttresses. Its square tower rises from the ground. Its yard and cemetery are enclosed by a rough stone wall. On down the road, a spring trickles from beneath green foliage. We string along. I am surprised by a railroad track. Michel says it is the line to Cherbourg. Laughter comes from under the bridge. Kids of all sizes are frolicking and splashing, having a great time. I envy their fun; it's like Rock Creek's Steel Bridge swimming hole when I was a youngster. We see an encampment of travel trailers and rubber tired wagons camped below the road in an open triangle, bounded by creek, field and road. Michel explains they are Gypsies. They camp here overnight at times. Four sorrel horses stand under trees near the creek; several men are lagging shot put size metal balls to a line drawn in a dirt lane.

Beatrice has assembled a delicious meal--*salmon pie, salad-- tomatoes, bread and cheese tray – ice cream and wine*--Her kitchen is next room beyond the dining – family room where we eat, visit, recall, and plan. Its fire place is huge and so is the round table. Michel tells Peggy about each cheese. We schedule morning's bakery trip before heading to bed.

Farm country, near the sea, is great for sleeping. It is not dawn patrol, however we are ready, more so when I fully understand breakfast is going to be here but comes from there. The little town is awake, the bakery is busy. We arrive home just as Chuck and Francois are raising the flags. François is accompanying with our Star Spangled Banner's tune, "La La La," all the way to the top--Daily raising and evening's lowering and the flags of our two countries has become a revered ceremony.

During *orange juice and coffee, croissants, brioche, and cakes, jam, and butter*, we discuss our day. Everyone goes. It takes two cars; we use no cell phones. Our first stop is a ruins of an old Romanesque church showing slab rocks placed in a slanted double row. Michel points them out in a wall, describing their positioning as "fish bone." I have difficulty catching them until Peggy interprets to me, "herring bone." Immediately we see. We were searching for fish fossils in rocks rather than the tilt of the rocks. Michel tells us this construction is found in eleventh century and some as far back as the ninth century.

Beatrice drives; Peggy and I thrill with the ride. We are in awe with her driving narrow roads between hedge rows and pointing simultaneously with both hands. On this day in Normandy "again," I am at times as scared as on my day "one" in 1944. We are traveling to Balleroy and Air Strip A-12.

We want to see the 362nd Fighter Group's memorial plaque located prominently in the Chapel on the grounds of the Balleroy Chateau It lists 362nd men lost here, similar to the plaques at Rennes and Prones. Balleroy Chateau is owned by the Malcolm Forbes family, gracious in displaying the plaque in their Chapel. It is in a frame on a white marble memorial alter. Their grounds and gardens are trimmed to perfection.

It is only a quarter mile back to town and a phone booth. Chuck admirably describes it as, "Beatrice hot-footing it for a phone booth." She is "footing" near a Memorial Monument mentioned in our 84 Return, with Peg's maternal grandmother's family name, Le Fevre. We drive a couple miles N. E. to the Romesnil Farm and our Chateaux Des Mogin's Maulers.

I feel much different than my return ten years ago. The orchard is still here but the "photo tree" with Chris, Don, and Red is gone. The boot scraper is in place. The light fixture at the comb of the roof is gone. Inside, the original wallpaper is still beautiful. Our reception is fantastic. Monsieur and Madame give our entourage the finest of mid-morning greetings with warm handshakes and seating for all at their classic rectangular table. Monsieur opens champagne and fine wine. Madame puts cakes on the table. Our time is pleasant. As we leave, I glance back at the cupboard door under the stairs. Once it was behind the bar from which after-mission liquor ration was dispensed.

We drive to A-12's pasture. An attractive white wooden fence is around the memorial. Beatrice has caused a fine marker to be placed at every American Ninth Air Force landing ground in Normandy. She leveraged sincere feelings with hard work and persuasion to mark all twenty-six Normandy airstrips. Each monument is unique and a work of art. Ours is a large native boulder tilted to the front, with an orange hued marble square plaque inscribed:

THE ADVANCED LANDING GROUND WAS BUILT
AT THIS SITE
BY THE NINTH AIR FORCE
JULY 17 1944 TO THE 10th OF AUGUST 1944
362nd FIGHTER GROUP

DEDICATED JUNE 13 1987

Cows are in pastures around the marker and across the road. Our three squadrons, about seventy-five planes, were parked next to bordering trees with dozed up dirt to form a revetment for each plane. Front line Sherman tanks were parked in the revetments when our Colonel arrived. The enemy was 1 1/2 miles away, and the same when we four new comers arrived on the 28th of July, 1944. Sure looks tranquil now. It takes time to remember; somehow we seem to hit "save" on our thoughts and move on, but not for long. Looking around and beyond to the borders of fields, something very subtle shows; it is part memory and part present. The trees have recovered but some show missing limbs, as though pruned by artillery. Battle damaged trees hit the recall button for the day we four rode in an open 6x6 truck to A-12 air strip. My mind's flicker-back is over quickly. The pace changes--Our photos now show trees and pastures with water troughs and dairy cows.

The folks in the lead car pull up for lunch at a cafeteria--great place--*ham loaf, potato salad, shrimp and a roll.* Peggy is given two little key ring animals as souvenirs, a pink rabbit and a green seal.

We walk dockside; see baskets of crabs and catches. Pont du Hoc by boat is our desire but seeing action similar to that of our own coastal towns is so interesting, we are late to the dock. We sweat out the line for tickets. Maybe they will sell out, full. Beatrice cools our apprehension, saying, "Voila" with a handful of tickets. She has foresight. We move into the boarding line, nearly last. I move to the top deck. This trip is guided in French so we don't get the nuances but there is no mistaking the tone. Chuck does video. We turn to starboard. In high school, I boxed a port-side opponent, and learned the difference between port and starboard. The sea is calm; I am standing high enough, so minor turns exaggerate the sway. We cruise along, maybe two km from shore. There is little beach, if any showing. The bluffs are more like steep little hills. This is not a description of how it is but of how I see it. We cruise for a half hour; things change as I ponder. I looked down on grounded ships the day we joined our squadron; now nothing, all is peaceful. The hill becomes a cliff, gets higher, rises rugged. We see the monument. We will visit the top land tomorrow.

Now, we have a much different point of view. The boat heads straight in, going right up close. I fantasize the past; speculate the planned routes, fire grapple hooks, climb the ropes. Trust buddies. My imagination fades into reality. I know so little about their job. I am reassured remembering the fine quality training given us for our flying jobs. I look forward to tomorrow's visit on top of this bluff, Ponte Du Hoc.

Returning to port, I admire the craft and its wake. I am proud of the French and United States flags flying at stern staffs. From my top perch, I frame them in churned white wake on blue water. Today, red shows only in the flags.

Looking at history, being with friends, foretells how we see and participate with the future.

Before we drive the short distance home, we make reservations for this evening's sea food dining. Beatrice wants to feed her dogs and give special care to the old one. She massages his limbs. He is large; she is tender, picks him up in a blanket and places him on the lawn in our conversation circle. He is able to walk a little when she takes him in.

Chuck and Fern have experienced this dinner. The restaurant is closed to-day. However, this afternoon our wonderful expediters cut red tape. *Voila* again, it will be done. We travelers host a seafood dinner Soirée. Each spectacular creation serves two people. A platter per pair is perfect. *All are attractive and heaped with crab, shrimp, oysters, crayfish, snails and unknowns. Bread, butter, condiments — lemon, vinegar, mayonnaise, shallots, and wine--all adorned with petite Allied flags on miniature poles.*

Evening at home, Beatrice shows Peggy the laundry beyond the kitchen with wood fired reservoir heater and mentions her attic drying room.

They invite me, too. Wow, the stairs feel just like they look, smooth and solid. The attic is the neatest storage place I have encountered. The floor is earthen, yes, dirt insulation about two inches deep. A hanging light with shade fits the atmosphere   Ancient items are too many to individually record in my brain. I ask if I may snap photos. I feel I am in a treasured room of family memories. Clothes lines stretch over attic's alleys — her drying room. The sun is low. I snap a picture through the single attic window overlooking the flags in the courtyard. My best photo is of a single item, an infant's wicker basket. Beatrice surprises us, saying, "All six babies used it." Wow again.

Tomorrow, we return our rental car to Cherbourg. Today has been full. Sleep finds me as soon as the pillow. I don't want to miss anything. Yesterday's events were a getting acquainted warm up. The bakery run is not necessary. Beatrice is in the kitchen, but leads us in raising the flags ceremony. This time, Chuck hums their Marseille as Beatrice and I raise the flags.

Beatrice's breakfast crepes are thin, browned to perfection and tasty. Crepes are special in our family, handed down to Peggy from her mother, from her little grandmother whose French father, Peggy's great grandfather, Antoine, liked his crepes pan-sized, thin, sprinkled with sugar and rolled-up. I am a winner. I know. I learned first breakfast, first morning after our wedding...Antoine knew his crepes.

We drive to Cherbourg and turn in our rental car. It will be a dawn departure tomorrow to catch our ferry, Cherbourg to Southampton, then train to Heathrow and fly to Oregon. We will "so-long" in England with Chuck and Fern. They leave Gatwick for Tennessee, but all that is tomorrow.

Francois wants us to see the Alabama Museum here in Cherbourg. We know nothing other than it has an American connection. We learn here and from sources at home: Alabama was built in England in 1862 and fitted in the Azores. In two years, the *C.S.S. Alabama,* famous Confederate Cruiser, sank, burned, or captured 66 Union ships. Union warships found Alabama in Cherbourg's harbor, June 1864, undergoing repair. The Alabama negotiated a battle, receiving three days to finish repairs, after which she sailed to meet the challenge of U.S.S. Kearsarge. Alabama was sunk. Most of the crew were rescued by the Kearsarge. However, Alabama's Captain and a few of the crew were rescued by a small ship which took them to England. I find the story fascinating.

This our last day, will be mostly of rediscovery. Yet like others, we are unscheduled and seldom pressed. Our goals are in sight. First, Ste-Me`re-Eglise, after lunch, back to Beatrice and Michel's, then to Omaha

Beach, Pointe du Hoc Monument and then the American Cemetery at Colleville-sur-Mer.

Yesterday, we looked straight up the route of the Rangers from the sea. Today, walking in the square and around the church in Ste-Me`re-Eglise, again brings speculative imagination, this time for the courage of the Airborne troops. Hanging from parachute shroud lines or climbing sheer bluffs bring back Col. Joe's words describing combat. I'm sure these men, too, were not afraid, maybe a little scared.

Back at their home, I feel privileged with a double invitation; Beatrice's to see her office, and Michel's, his shop. They are across the yard past the flags in the long building. Michel has a map for me. Note – "He has done remarkable wood work and has beautiful power equipment."

I wrote on back of her photo, "Beatrice's office has books, maps, sketches, drawings-9th Air Force material. Her projects show this is a working office of a great lady who told me she didn't care for Paris because everyone ran all the time." (I never saw her longer than 30 seconds when she idled inactive.)--"Drives like a demon, but with skill. She is a wonderful little lady with tremendous presence and is straight forward. She is quick, intense, gracious, and paints her toenails bright red."

Again on the go, we head for Omaha Beach. In a small village we see a simple and most direct message on a banner above the street. It was flanked with the Allies' flags:

6 JUNE 44   THANK YOU
6 JUNE 94   WELCOME

Omaha beach is even more tranquil than in nineteen eighty-four; then, we were coming back to scenes of forty years before. Instead of twenty-one, I was sixty-one and walking on these sands for the first time.

My vivid 1944 memory is looking down across crowded approaches through both sea and land. We were seeing it as four replacement pilots, with gear loosely filling the cargo floor in a 9th Air Force C-47, banking for a landing. The war battered road from the delivery airstrip to our fighter airstrip was lined with mangled materiel.

Thinking of those attacking D-Day men, and the toll, causes my thoughts to focus again on those dreadful hours on Omaha beach, when eventually the exposed, skilled and brave rallied their broken units into a do or die success.

I have read of D-Day, especially the roles of strategic and tactical planning. Second guessing is easy. When I think of this landing, I ask what would have made it better on Omaha Beach.

My questioning was stimulated by an irrigation equipment representative. I was in my late fifties, he early thirties. I had experience in irrigation application and water management and thought myself coachable. I don't want to agonize over why his analogy was appropriate. He brought out a cartoon showing two armies of ancient times obviously preparing to clash. Pennants are flying; a General in Roman battle dress stands, looking beyond his ranks, surveying the field of imminent conflict. Hordes of opposition stand ready with swords and spears for the charge. A young salesman stands at the corner of the General's field tent, pleading for the General's attention. Out of sight, behind the General's headquarters tent, the salesman has a tripod mounted Gatling gun. The cartoon's caption reads:

"Don't bother me now. Can't you see I have a battle to fight?"

I have pondered why fighter bombers were not used to knock out the enemy on the dunes above.

We stop at the Omaha Beach Landing monument. The beach has few people to-day. It is clean and smooth. A long wooden object, maybe a boat, is washed broadside, far-out, at the surf line. I can imagine it being Norseman, looked like a log at first glance. We make our way on and up to Ponte du Hoc.

It isn't crowded; the plaque is modeled like an open book; English on the left, French on the right. Hushed viewers make room for a photo of the plaque and spire monument.

TO THE HEROIC RANGER COMANDOES
D2RN E2RN F2RN
OF THE 116TH INF
WHO UNDER THE COMMAND OF
COLONEL JAMES E. RUDDER

Beatrice's Normandy

OF THE FIRST AMERICAN DIVISION ATTACKED AND TOOK POSSESSION OF THE PONT DU HOC

A large expanse of bench-lands is here above the beach. The area shows heavy pounding by bombardment of both bombs and Naval shells. Paved foot paths meander among deep holes. Beatrice tells me, "They leveled most of the craters when establishing the foot paths." She added, "It was solid shell and bomb craters on top of craters."

The remaining German gun fortifications are massive, with openings similar in size to our low level targets on the strafing ranges at Luke's Gila Bend and Baton Rouge and again at the Welsh coast target range in England.

We walk the paths, look into fortifications. I remember our feeling when shelled and bombed on our Normandy air strip. Shelling came day or night. We hunkered down in fox hole/slit trenches and survived.

Combat pilots from North Africa, Sicily, and Italy had rotated to become instructors. Our fighter training squadron at Harding Field, Baton Rouge was commanded and led by First Lieutenant Schranz. Fighter Bomber technique was honed to carry out precise low level attacks. Allies had fought German General Rommel on his way out of Africa. Again, he is the opposition. We want to push him from Normandy.

I cannot remember anything more fierce than strafing fighters coming at you, their wings lit up by blazing 50 caliber tracers. It is much worse--four more unseen bullets are coming in for each one visible; a two ship element, sixteen guns, 8 + rounds per second for each gun, 130 rounds per second. Multiply by 2-4 since each pass has multiple trigger bursts, call it about 400 rounds coming as, thick as hailstones. It's enough to rattle through slots and cracks, blow up ammo stores, while ricochets strike like angry hornets. Outside in open pits, enemy air and beach weapons of 30 and 50 caliber, and 20s, 40s and 88 mm guns were a Thunderbolt specialty.

I read the D-Day 50th Anniversary handouts; a Michelin 1939 Normandy map with overlay of the five invasion beaches; it briefly describes all five beaches' landings. The Normandy Landings 6 June

1944/United States Commemorative Activities 5-6 June 1994, a booklet with photos and maps detailing invasion units and description of the Army, Navy, and Air force participation; it expands including Rangers, Paratroops and Glider units:

"At Omaha beach...Captain Harry Sanders, commanding Destroyer Division,18 took his ships literally into the surf, as little as 800 yards from shore, to blast stubborn defenders from the hills above...General Leonard T. Gerow, acknowledged Sanders' courage in a message to Admiral Kirk: "Thank God for the Navy."

Thank God for the Navy who ad-libbed and did arrive, after hours of ineffective shelling but remember the men who struggled off landing craft, to be immediately be swamped or pinned down by enemy fire from the dunes above. Troops landing with their amphibious tanks floundering proved a difficult force to either enable or defend by warships alone.

"Troops of Omaha were pinned down among debris, and if lucky, close behind the sea wall, while the less deadly beaches, American Utah, Canadian Juno, and British, Gold and Sword, were up and across the sand, taking cover and territory."

The D-Day map synopsis summed this way:
"At Omaha, the Americans encountered grave difficulties and at sunset were still under German artillery fire; troops from Utah managed to make contact with the Airborne Forces."

From 362nd GP; 377th, 378th, and 379th squadrons: This came from the 379th report of the group's briefing: ...*The most important event took place when our group was given its assignment. <u>We found ourselves to be a 'Readiness Group' and could be called upon for any kind of mission</u>. The pilots were told of the alert system. The squadron (379th) was to be on the runway, one squadron to be on fifteen minute alert, the other on one hour alert.*

The next morning, June 6th, our Squadron was out bright and early; no word of a mission had been received when General Eisenhower announced the much waited, "The Invasion started at six-thirty this morning." <u>No mission came in all that day</u>, but we drew a hard assignment for that night--we were to take off and escort C-47s pulling *gliders to the beachhead. At first we thought the mission would be a hard one but on the return the pilots said nothing much had happened.*

A squadron of four flights of four ships--one flight for cover--three flights strafing, a two-ship element at a time could enable troops to cross the sands, and tip Omaha's invasion field our way. Casualties would have reduced hours sooner.

Patrolling squadrons of fighter planes were assigned only to prevent enemy fighters' strafing attacks on landings. Communication through a multiple chain of command didn't work. Reaction was slow. In boxing terms, late and missed haymakers alone fail. Needed are hooks and jabs to rock the opponent on his heels. In this case, get our men moving, clear the beach, counter punch--hit him in his face, close an eye, lets troops win ground.

Our Generals and Admirals likely had similar concerns when experiencing the wrong end of fighter strafing attacks; so much so, the terms--air support, air cover, and ground support were used synonymously to describe preventing enemy aircraft (e/a) getting through to strafe or bomb our units on the ground. In short order, after Omaha, the word "support" switched its meaning from defense to offense. Thunderbolts both defended and enabled troops and their armor to move forward.

The lessons learned paid dividends a little later with the "breakout" and fast thrust across France.

Our day is going very well. Somewhere back in our minds, we know this is the last of reconnect time. These days of being with Beatrice and her family in a country where our people joined in righting a wrongful use of power is meaningful for us. Today we learn the price.

THE AMERICAN BATTLE MONUMENTS COMMISSION:

"Normandy Cemetery is situated on a cliff overlooking Omaha Beach and the English Channel...in Colleville-sur-Mer...covers 172½ acres and contains the graves of 9,386 of our military Dead, most of whom gave their lives in the landings and ensuing operations. On the walls of the semicircular garden on the east side of the memorial are inscribed the names of 1,557 of our Missing who rest in unknown graves."

Today, few people are visiting as compared to eleven days ago when we joined several thousand families, friends and comrades in commemorative respect. We walk. I can't help noting the alignment of rows of markers so precise. Any angle I look from where I stand, they radiate out, straight and beyond my comprehension. Headstones are white crosses interspersed here and there with white Star of David markers.

Epilogue: Connecting Then to Now

Peg and I reminisce with each word. WWII remains. In the last twenty-five years we have bonded with new and old friends and their families from our distant past. The 362nd Fighter Group, Ninth Air Force, holds us. The tethers are long. Thoughts appear, flicker then light, while we live in bonus time.

We have used a quarter ton of paper, fifty yards of printer ribbon, and quarts, a spoon at a time, of ink in cartridges.

There are eight large three ring binders of text: Two binders with Mission Reports and filing notes. Cases of volumes of war history, two wartime scrapbooks, and thousands of photos gathered then and during four returns.

I've whittled a hundred walking sticks (keeps me awake) in the last three years while Peggy reads to us; not just war, but geology, anthropology, humor, tragedy, North Pole, Antarctica, and politics. Roughly, a book a stick.

We have traveled a little. Mostly in the US, but also to England, France, Germany, Finland, St. Petersburg and Siberia in Russia, and western Canada, with camera as note pad, sometimes video, and recorder.

Our file boxes and book shelves are like oxygen is to breathing. It's there. We use so very little, but benefit so very much by its presence. Like harvesting a garden or watching a cloud, it's ours; we have it. It is a part of and precious to us.

Listening and talking and knowing a little about our comrades and families, all is valuable; maybe only a moment. I'll jot down a couple words.

Battle Cats was written after we attended the 362nd's 1992 dedication in Wormingford, England.

The 362nd Fighter Group, nicknamed Moggin's Maulers, flew from two bases in England, a forward Normandy airstrip, two more strips and a base in France, finally from two airfields in Germany. The group entered combat in February 1944 and fought through to the end, in May of 1945.

I and three cadet-days friends were among the first replacement pilots to join the outfit in Normandy. We were shelled in our first half hour, flew combat second day, lived in foxholes, were shot at in the traffic pattern and bombed nightly. Ground personnel were magnificent; worked like beavers, repaired every break, kept the planes fixed, fit and flying. Two were killed by shell fire at our Normandy airstrip A-12.

Like fierce battle cats perched on the mat-covered shoulder of their Asian warriors in hand to hand combat, we too, slashed scratched, snarled and bit for our warrior, General George S. Patton's Third Army. We covered his columns in their drive across France. We protected his flanks, pinned down his opposition, blew up locomotives, freight and tank cars, bridges and tunnels; turned German ammo and fuel dumps into volcanic spectaculars. We destroyed swastika bearing Buzz bombs, ME-109s, FW-190s, staff cars, halftracks, artillery and 88, 40, and 20 mm anti-aircraft guns. But the thing we were most appreciated for was knocking out enemy armor. The German 88 mm gun on a Tiger Tank could hold a column at bay and rapier it to oblivion.

Our tanks were outgunned by "Jerry" until effective tactical teamwork evolved from the dust and death of Normandy's battle duels. P-47 combat pilots on the ground in forward armored units directed their Thunderbolt squadrons to troubling targets.

We bristled with two five-hundred pound bombs, sometimes carried two one- thousand pounders; four five-inch rockets, and eight 50 caliber machine guns. Our fighters were Patton's stiletto, we worked up close.

Our 377th squadron's insignia is a Cougar snarling from a cloud. 'Battle cats' riding on shoulders fit our mission tooth and claw. The huge fighter plane showed the ferocity of a cat with spine arched; its young pilots had the will to inflict the threat.